Materials and Life Science Experimental Facility (MLF) at the Japan Proton Accelerator Research Complex (J-PARC)

Materials and Life Science Experimental Facility (MLF) at the Japan Proton Accelerator Research Complex (J-PARC)

Topical Collection Editor
Klaus-Dieter Liss

MDPI • Basel • Beijing • Wuhan • Barcelona • Belgrade

MDPI

Topical Collection Editor
Klaus-Dieter Liss
Guangdong Technion—Israel Institute of Technology (GTIIT)
China

Editorial Office
MDPI
St. Alban-Anlage 66
4052 Basel, Switzerland

This is a reprint of articles from the Topical Collection *Facilitices* published online in the open access journal *Quantum Beam Science* (ISSN 2412-382X) from 2017 to 2018 (available at: https://www.mdpi.com/journal/qubs/special_issues/facilities)

For citation purposes, cite each article independently as indicated on the article page online and as indicated below:

LastName, A.A.; LastName, B.B.; LastName, C.C. Article Title. *Journal Name* **Year**, *Article Number, Page Range.*

ISBN 978-3-03897-483-3 (Pbk)
ISBN 978-3-03897-484-0 (PDF)

Cover image courtesy of Klaus-Dieter Liss.

Contents

About the Topical Collection Editor vii

Preface to "Materials and Life Science Experimental Facility (MLF) at the Japan Proton
Accelerator Research Complex (J-PARC)" ix

Klaus-Dieter Liss
Materials and Life Science with Quantum Beams at the Japan Proton Accelerator
Research Complex
Reprinted from: *Quantum Beam Sci.* **2018**, *2*, , doi:10.3390/qubs2020010 1

Hiroshi Takada, Katsuhiro Haga, Makoto Teshigawara, Tomokazu Aso, Shin-Ichiro Meigo,
Hiroyuki Kogawa, Takashi Naoe, Takashi Wakui, Motoki Ooi, Masahide Harada and
Masatoshi Futakawa
Materials and Life Science Experimental Facility at the Japan Proton Accelerator Research
Complex I: Pulsed Spallation Neutron Source
Reprinted from: *Quantum Beam Sci.* **2017**, *1*, 8, doi:10.3390/qubs1020008 7

Kenji Nakajima, Yukinobu Kawakita, Shinichi Itoh, Jun Abe, Kazuya Aizawa,
Hiroyuki Aoki, Hitoshi Endo, Masaki Fujita, Kenichi Funakoshi, Wu Gong, et al.
Materials and Life Science Experimental Facility (MLF) at the Japan Proton Accelerator Research
Complex II: Neutron Scattering Instruments
Reprinted from: *Quantum Beam Sci.* **2017**, *1*, 9, doi:10.3390/qubs1030009 33

Kaoru Sakasai, Setsuo Satoh, Tomohiro Seya, Tatsuya Nakamura, Kentaro Toh, Hideshi
Yamagishi, Kazuhiko Soyama, Dai Yamazaki, Ryuji Maruyama, Takayuki Oku, et al.
Materials and Life Science Experimental Facility at the Japan Proton Accelerator Research
Complex III: Neutron Devices and Computational and Sample Environments
Reprinted from: *Quantum Beam Sci.* **2017**, *1*, 10, doi:10.3390/qubs1020010 92

Wataru Higemoto, Ryosuke Kadono, Naritoshi Kawamura, Akihiro Koda, Kenji M. Kojima,
Shunsuke Makimura, Shiro Matoba, Yasuhiro Miyake, Koichiro Shimomura and
Patrick Strasser
Materials and Life Science Experimental Facility at the Japan Proton Accelerator Research
Complex IV: The Muon Facility
Reprinted from: *Quantum Beam Sci.* **2017**, *1*, 11, doi:10.3390/qubs1010011 127

About the Topical Collection Editor

Klaus-Dieter Liss is a Professor at the Guangdong Technion–Israel Institute of Technology in Shantou, China (since 2017), holding an academic appointment as Full Professor at the Technion, Israel. He also is an Honorary Professor at the school of Mechanical, Materials, Mechatronic, and Biomedical Engineering at the University of Wollongong, Australia (since 2014). Starting in 1983, he studied General Physics at the Technische Universität München, Germany (Dipl. Phys., 1990) and defended a thesis at the RWTH Aachen, Germany (Dr. rer. nat., 1995) supported by a doctoral scholarship at the Institut Laue-Langevin (ILL), France. His career in neutron-scattering started in 1986, spending continuously several months per year at the ILL, complemented by high-energy synchrotron radiation since 1991. In 1994, Liss started as a postdoctor at the European Synchrotron Radiation Facility (ESRF) in France, followed by a full position as beamline scientist and beamline responsible from 1995 to 2001. He took one semester sabbatical leave, lecturing diffraction theories at the Friedrich-Alexander-Universität Erlangen-Nürnberg, Germany in 1998–1999. After a short time in Germany at GKSS Research Center, DESY and Technische Universität Hamburg Harburg (2001–2003), he moved as a Senior Researcher to the Bragg Institute of the Australian Nuclear Science and Technology Organisation, ANSTO in 2004, where he enjoyed the inaugural Senior Research Fellowship (2007–2010) of the organization. In 2013–2014 he spent a sabbatical year at the Japan Atomic Energy Agency in Tokai, and he is a regular user of J-PARC. In the recent years, Liss' research has focused on neutron and synchrotron diffraction methods for the investigation of thermo-mechanical processes, mainly on metals, including in situ and time-resolved measurements and pioneering experiments to study phase transformations, microstructure evolution, order and disorder, and defect kinetics. Current research themes encompass materials processing under extreme conditions and in-situ investigations, fast and ultrafast time responses in materials, and novel and enhanced quantum beam sources for application in materials science and physics. Klaus-Dieter Liss held presidency of the Materials Australia New South Wales Branch (2014–2017) and was National Councilor in that society (2011–2017). He is Fellow of the Australian Institute of Physics and Member of the TMS Phase Transformations Committee. As Editor-In-Chief, he created the journal Quantum Beam Science and is member of the Editorial Board of Metals, as well as Advanced Engineering Materials. He has been Guest Editor for MRS Bulletin, June 2016, and for MRS Proceedings. In 2012 and 2014, Liss chair-organized the MRS Fall Meeting symposia on applications of neutron and synchrotron radiation, reading tutorials, as well as at TMS. He organized similar symposia at PRICM-8 (2013) and THERMEC (2011, 2013, 2016, and 2018).

Preface to "Materials and Life Science Experimental Facility (MLF) at the Japan Proton Accelerator Research Complex (J-PARC)"

Welcome to the booklet *Materials and Life Science Experimental Facility (MLF) at the Japan Proton Accelerator Research Complex (J PARC)*, which is the first special edition of Quantum Beam Science and selected from its Topical Collection *Facilities*. It compiles four in-depth technical articles describing design and capabilities of this large-user-facility, rounded up by an Editorial, and serving as a reference for their users and the scientific community. As Editor-In-Chief, I like to thank in-depth the Editorial Board Members who have supported these articles, particularly Prof. Dr. Toshiji Kanaya, Head of MLF at J PARC.

Klaus-Dieter Liss
Topical Collection Editor

quantum beam science

MDPI

Editorial

Materials and Life Science with Quantum Beams at the Japan Proton Accelerator Research Complex

Klaus-Dieter Liss [1,2]

1   Guangdong Technion — Israel Institute of Technology, Shantou 515063, China; kdl@gtiit.edu.cn or
    kdliss@technion.ac.il or liss@kdliss.de; Tel.: +86-0754-8807-7102
2   Technion — Israel Institute of Technology, Haifa 32000, Israel

Received: 7 May 2018; Accepted: 14 May 2018; Published: 14 May 2018

1. Introduction

The Materials and Life Science Experimental Facility (MLF) is the major neutron and muon quantum beam facility in the eastern part of the world. Besides the neutrino experiment and the hadron facility, it forms part of the Japan Proton Accelerator Research Complex (J-PARC), a scientific landmark installation pictured in Figure 1. The MLF delivers quantum beams to 20 neutron beamlines, plus one under commissioning, and two muon beamlines hosting three end-stations. Figure 2 displays a schematic layout, and an impressive view into one of the experimental halls is shown in Figure 3. The first neutron and muon beams were produced in 2008. Nowadays, the facility serves about 700 peer-reviewed and accepted proposals per year by users coming from 34 countries across the globe. Applications range from engineering to life science, include soft and hard condensed matter and consider cryogenic conditions, high temperatures, and extreme pressure. MLF is a division of J-PARC governed by a consortium of the Japan Atomic Energy Agency (JAEA) [1] and the High Energy Accelerator Research Organization, KEK [2], as stakeholders. User promotion services and operation of the seven MLF public-access facilities are codirected by the Comprehensive Research Organization for Science and Society (CROSS) [3], not only cooperating closely with the J-PARC Users Office but also providing supporting scientific manpower, safety and scientific management.

The book on *Materials and Life Science with Quantum Beams at the Japan Proton Accelerator Research Complex* is an extract of the larger Topical Collection *Facilities* in the journal *Quantum Beam Science* [4], compiling a comprehensive description of MLF at J-PARC. Interestingly, the concept of Quantum Beam Science reunifying various short-wavelength and high-energetic electromagnetic and particle beams, predominantly for applications in materials and life sciences [5], has been minted in Japan around the stakeholder organizations [6]. It strongly recognizes the need for various quantum beams for the investigation of peculiar materials science problem. Accordingly, MLF operates two major quantum beam sources—neutrons and muons. Complementary access at Japan's major synchrotron SPring-8 and other large-user facilities is strongly supported in the beam-time application and reviewing processes.

2. Overview and Chapters

The book has been separated into four review articles, as published in the journal *Quantum Beam Science* [8–11]. Central to all J-PARC experimental divisions is the landmark accelerator complex (Figure 1), composed of a linear accelerator (LINAC) feeding H^- ions by stripping them to protons into the Rapid Cycle Synchrotron. Further accelerated to 3 GeV, those protons are used to produce the muon and neutron sources at MLF. Besides feeding MLF, protons can be further accelerated to 50 GeV in the main accelerator ring, serving the Hadron Beam Facility for creating pions, kaons and other hadrons for fundamental nuclear and particle physics, including the higher generation quark families, such as the strange and charm quarks [12]. The neutrino facility is another experiment of global character, producing light-mass particle radiation in an underground station with detectors both nearby and 295 km across Honshu island at the Super-Kamiokande detector, chasing muon–neutrino oscillations to probe the most fundamental coupling constants of the universe [13]. The scope of the present book, however, focuses on MLF and its neutron and muon beams as applied to a vast range of disciplines based on the material and life sciences. Their central layout is given in Figure 2, with respective sources and beamlines, occupying buildings of hundred-meter dimensions and vast experimental halls, Figure 3.

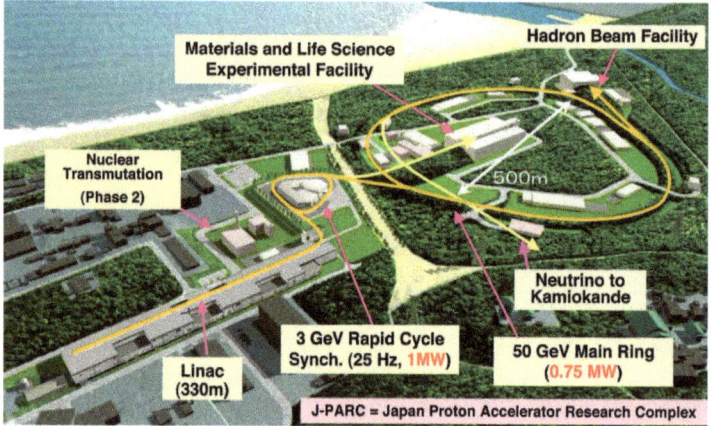

Figure 1. Schematic aerial view of the entire J-PARC complex. To the left is the linear accelerator (LINAC), feeding protons into the rapid cycle synchrotron. Once accelerated to 3 GeV, they are delivered to the Materials and Life Sciences facility, or alternatively further accelerated to 50 GeV and delivered to one of the other two large laboratories: the hadron facility and the neutrino experiment. (Modified from "The entire view of J-PARC" by Shoji Nagamiya and published under a CC BY-SA 3.0 license [7]).

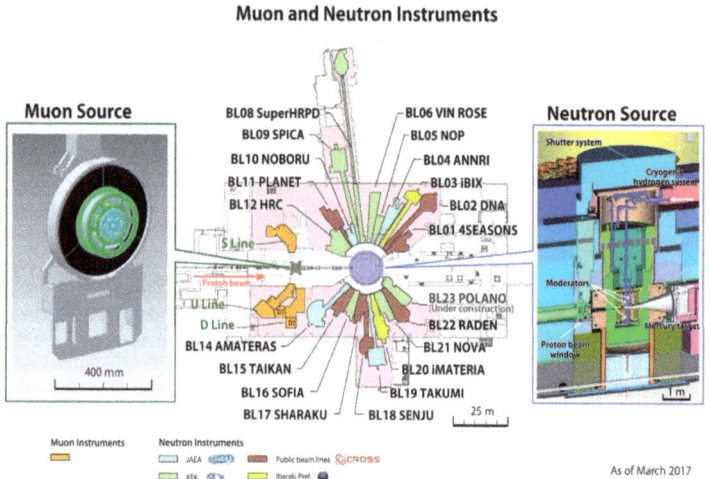

Figure 2. Layout of the muon and neutron beamlines (**center**) with drawings of the sources (**left and right**). The color codes of the neutron beamlines denote the operator organizations as per legend. (Reproduced from [14] with permission). JAEA: Japan Atomic Energy Agency; KEK: High Energy Accelerator Research Organization; CROSS: Comprehensive Research Organization for Science and Society.

Figure 3. Photograph of the Materials and Life Science Experimental Facility (MLF) Experimental Hall II at J-PARC with the muon D-line shielding in the foreground and looking over the neutron beamlines BL14 in the image center to BL23 at the far back (compare to Figure 2).

The four sections are divided into descriptions of the Pulsed Spallation Neutron Source [8], the Neutron Scattering Instruments [9], the Neutron Devices and Computational and Sample Environments [10], and the Muon Facility [11].

2.1. Pulsed Spallation Source

The core of the neutron beam lines is the pulsed spallation source reviewed in the article by Takada et al. [8]. The target is the device where the accelerated proton beam hits heavy atoms, and spallation takes place by excitation of the nuclei and subsequent evaporation of neutrons. Moreover, time-of-flight neutron analysis is used throughout the beamline facilities, requiring well-defined neutron pulses. Spallation at the liquid mercury target produces high-energy neutrons, which need to be moderated into the desired thermal and cold neutron spectral distributions. Proton beam pulse shape, the spallation process, location and optics of the moderators, and type of instrument determine the neutron pulse shape, which is essential to optimize instrument resolution. The proton beam power is designed up to 1 MW, rendering the engineering technology of the target and moderators critical for handling mechanical shock waves, cryogenic cooling, and radiation damage. The review describes in detail such engineering solutions and the performance of the beams.

2.2. Neutron Scattering Instruments

The main focus of a user of the neutron facility is on the kind and capabilities of beamlines and neutron scattering instruments, reviewed here by Nakajima et al. [9]. The 21 dedicated and unique beamlines are grouped according to their methods and applications into (i) Spectroscopy, (ii) Crystal-Structure, (iii) Nano-Structure and (iv) Pulsed Neutron Application.

Crystal structures (ii) are determined by the neutron diffraction method, obeying elastic momentum transfer by Bragg's law or the Laue equation. The differences and peculiarities of the seven beamlines lie in extra-high resolution, and sample environments for classes of materials—such as engineering, chemistry, biological crystals, or specimens under extreme conditions.

On top of a potential diffraction component, spectroscopy (i) deals with inelastic scattering processes where the neutron loses or gains energy due to interaction with lattice vibrations—phonons, diffusion, potential-barrier tunneling, or other dynamic and kinetic processes occurring in the specimen. As with the diffraction group, the six instruments vary according to applications and specifications. The latest instrument under commissioning is a polarized neutron spectrometer, sensitive to magnetic and spin excitations in solid materials.

Objects larger than atomic and basic crystal structures, such as micelles, cells, precipitates, polymers, lipids, foams, hierarchical structures, domains, quantum dots, amorphous materials, thin films, etc. scatter on the 'small angle' or 'small momentum transfer' small-Q scale, forming the nano-structure group (iii) by probing nanometer length scales. The four dedicated beamlines extend over small-Q bulk scattering, (polarized) reflectometry for (magnetic) thin film and interface studies as well as instruments covering the overlap from small-Q to the conventional diffraction regime, and the measurement of pair distribution functions.

Finally, fundamental neutron physics, neutron optics and the important field of neutron imaging are grouped into the pulsed neutron applications (iv). Here the instrumentation takes particular advantage of the pulsed neutron bunches, such as the recent technique development of energy-resolved neutron imaging, where Bragg edge diffraction is used in transmission radiography to obtain or discriminate the signal by crystallographic information and obtain additional contrast.

Altogether, it can be stated that J-PARC/MLF has developed and constructed world-class, state-of-the-art beamlines covering all aspects of neutron scattering with respect to complementary and combined, multi-dimensional or time-resolved data acquisition, serving a variety of disciplines and sample systems.

2.3. Neutron Devices and Computational and Sample Environments

The aforementioned beamlines, coverage of disciplines, length and time scales, specimen conditions and multi-dimensional acquisition methods require technological development of devices, computational data acquisition and instrument control, and special sample environments, described here by Sakasai et al. [10]. There exist a multitude of optical and timing devices—such as focusing mirrors, neutron guides, choppers at one end—matched by a scattering geometry with sample handling and a secondary spectrometer to the detector. As each beamline is unique with very specific requirements, so are those devices.

Each beamline may consist of hundreds to thousands of detector channels of which each may be divided into a neutron time-of-flight pattern; each detector angle may be different to its neighbor pixel, recording different time-of-flight patterns that finally have to be calibrated and mapped into physically meaningful, sample-specific diffractograms and energy-transfer spectra. The measured dimensions can be high as three-dimensional reciprocal space plus energy resolution, represented in four dimensions. Similar dimensionality or higher is obtained in three-dimensional computed tomography with reciprocal space and time information in addition.

MLF instruments are among the first neutron scattering facilities to have implemented so called event-mode, in which a time stamp for each registered neutron is recorded, allowing to re-bin histograms after the experiment from the stored data. Such practice necessitates not only huge storage capacities but also computing algorithms and user interfaces, which have been developed and are presented.

Last but not least, although special sample environments such as high-pressure devices were described in the previous chapter, their electronics and integration demands special considerations, as reported in the same chapter.

2.4. The Muon Facility

The chapter by Higemoto et al. [11] describes the muon facility, which at first sight appears distinct from the neutron facility, however it is complementary for some user communities, e.g., for the determination of magnetic structure and properties of a material. Not only do the operation of the sources show synergies, but there are also similarities in data acquisition systems and infrastructure. Since the user community and physical installation are much smaller than for neutron scattering, the article reviews all aspects from the source, beamlines and applications. Similarly, the flux and energy of muons vary. For example, in conventional spectroscopy where a muon is embedded in the bulk of a material, its decay time and behavior reveal atomic local structural properties. Or moreover, the conditioning of ultra-slow muons with particular low kinetic energies can specifically probe the surface of a specimen. Examples of applications are reviewed on a new antiferromagnetic phase in iron-based superconductor, magnetic ground states of iridium spinel compound, or non-invasive element analysis by muonic X-ray spectroscopy.

3. Conclusions and Outlook

With a unique park of instrumentation and technological development, 10 years after first beam MLF at J-PARC is one of the world-leading facilities, serving over 700 proposals per year. It is one of the three major spallation neutron sources operating and one of the four muon facilities worldwide, and has potential for further development in both beam power and instrumentation. While the last neutron beamline slots are being occupied and commissioned, effort is focused on dedicated sample environments which are second to none in the world.

I would like to express my congratulations to the facility and invite the readers and experts to enjoy the chapters of the present book on the J-PARC Materials and Life Science Experimental Facility for reference, documentation and support of their research and development.

Conflicts of Interest: The author declares no conflict of interest.

References

1. Japan Atomic Energy Agency. Available online: https://www.jaea.go.jp/english/ (accessed on 30 April 2018).
2. KEK, High Energy Accelerator Research Organization. Available online: http://www.kek.jp/en/ (accessed on 30 April 2018).
3. CROSS Neutron Science and Technology Center. Available online: https://neutron.cross.or.jp/en (accessed on 30 April 2018).
4. Liss, K.-D. Facilities in Quantum Beam Science. *Quantum Beam Sci.* **2018**, *2*, 6. [CrossRef]
5. Liss, K.-D. Quantum Beam Science—Applications to Probe or Influence Matter and Materials. *Quantum Beam Sci.* **2017**, *1*, 1. [CrossRef]
6. Yamamoto, H.; Igawa, N.; Moriai, A.; Sakashita, T.; Ohba, H.; Sekiguchi, T.; Yasuda, R.; Kawano, T.; Suzuki, E. (Eds.) *Annual Report QuBS 2013*; Quantum Beam Science Directorate, Japan Atomic Energy Agency: Tokai, Japan, 2014.
7. Nagamiya, S. Introduction to J-PARC. *Prog. Theor. Exp. Phys.* **2012**, *2012*, 020001. [CrossRef]
8. Takada, H.; Haga, K.; Teshigawara, M.; Aso, T.; Meigo, S.-I.; Kogawa, H.; Naoe, T.; Wakui, T.; Ooi, M.; Harada, M.; et al. Materials and Life Science Experimental Facility at the Japan Proton Accelerator Research Complex I: Pulsed Spallation Neutron Source. *Quantum Beam Sci.* **2017**, *1*, 8. [CrossRef]
9. Nakajima, K.; Kawakita, Y.; Itoh, S.; Abe, J.; Aizawa, K.; Aoki, H.; Endo, H.; Fujita, M.; Funakoshi, K.; Gong, W.; et al. Materials and Life Science Experimental Facility (MLF) at the Japan Proton Accelerator Research Complex II: Neutron Scattering Instruments. *Quantum Beam Sci.* **2017**, *1*, 9. [CrossRef]
10. Sakasai, K.; Satoh, S.; Seya, T.; Nakamura, T.; Toh, K.; Yamagishi, H.; Soyama, K.; Yamazaki, D.; Maruyama, R.; Oku, T.; et al. Materials and Life Science Experimental Facility at the Japan Proton Accelerator Research Complex III: Neutron Devices and Computational and Sample Environments. *Quantum Beam Sci.* **2017**, *1*, 10. [CrossRef]
11. Higemoto, W.; Kadono, R.; Kawamura, N.; Koda, A.; Kojima, K.M.; Makimura, S.; Matoba, S.; Miyake, Y.; Shimomura, K.; Strasser, P. Materials and Life Science Experimental Facility at the Japan Proton Accelerator Research Complex IV: The Muon Facility. *Quantum Beam Sci.* **2017**, *1*, 11. [CrossRef]
12. Noumi, H. Strange and Charm Hadron Physics at J-PARC in Future. In Proceedings of the 12th International Conference on Hypernuclear and Strange Particle Physics (HYP2015), Sendai, Japan, 7–12 September 2015.

13. Abe, K.; Abgrall, N.; Aihara, H.; Ajima, Y.; Albert, J.B.; Allan, D.; Amaudruz, P.-A.; Andreopoulos, C.; Andrieu, B.; Anerella, M.D.; et al. The T2K experiment. *Nucl. Instrum. Methods Phys. Res. Sect. A Accel. Spectrom. Detect. Assoc. Equip.* **2011**, *659*, 106–135. [CrossRef]
14. Kojima, K.M.; Kawamura, N.; Mishima, K.; Naoe, T.; Oikawa, K.; Parker, J.; Sakasai, K.; Tominaga, T.; Tanaka, K. (Eds.) *J-PARC MLF Annual Report 2016*; J-PARC: Tokai, Japan, 2017.

quantum beam science

MDPI

Review

Materials and Life Science Experimental Facility at the Japan Proton Accelerator Research Complex I: Pulsed Spallation Neutron Source

Hiroshi Takada *, Katsuhiro Haga, Makoto Teshigawara, Tomokazu Aso, Shin-Ichiro Meigo, Hiroyuki Kogawa, Takashi Naoe, Takashi Wakui, Motoki Ooi, Masahide Harada and Masatoshi Futakawa

J-PARC Center, Japan Atomic Energy Agency, Tokai, Ibaraki 319-1195, Japan; haga.katsuhiro@jaea.go.jp (K.H.); teshigawara.makoto@jaea.go.jp (M.T.); aso.tomokazu@jaea.go.jp (T.A.); meigo.shinichiro@jaea.go.jp (S.-I.M.); kogawa.hiroyuki@jaea.go.jp (H.K.); naoe.takashi@jaea.go.jp (T.N.); wakui.takashi@jaea.go.jp (T.W.); ohi.motoki@jaea.go.jp (M.O.); harada.masahide@jaea.go.jp (M.H.); futakawa.masatoshi@jaea.go.jp (M.F.)
* Correspondence: takada.hiroshi90@jaea.go.jp; Tel.: +81-29-282-6424

Received: 9 May 2017; Accepted: 20 July 2017; Published: 2 August 2017

Abstract: At the Japan Proton Accelerator Research Complex (J-PARC), a pulsed spallation neutron source provides neutrons with high intensity and narrow pulse width pulse to promote researches on a variety of science in the Materials and Life Science Experimental Facility (MLF). It was designed to be driven by a proton beam with an energy of 3 GeV, a power of 1 MW at a repetition rate of 25 Hz, that is world's highest power level. It is still on the way towards the goal to accomplish the operation with a 1 MW proton beam. In this review, distinctive features of the target-moderator-reflector system of the pulsed spallation neutron source are presented.

Keywords: spallation neutron source; mercury target; moderator; para-hydrogen; cavitation; pressure waves; microbubbles; cryogenic hydrogen system; 3 GeV proton beam transport

1. Introduction

The Japan Proton Accelerator Research Complex (J-PARC) is a multi-purpose research facility complex consisting of 3 accelerators and 4 experimental facilities. The Materials and Life science experimental Facility (MLF) is one of experimental facilities of J-PARC, having muon and neutron facilities. At MLF, a pulsed spallation neutron source was built to provide high-intensity and high-quality neutron beams for cutting-edge research on a variety of materials science [1]. It was designed to receive a 3 GeV proton beam with 333 μA, the power of 1 MW, at a repetition rate of 25 Hz. The proton beam is accelerated up to 0.4 GeV by a linac and accumulated in 2 short bunches, and then accelerated up to 3 GeV in the Rapid Cycling Synchrotron (RCS), resulting to have a structure of a 150 ns bunch width with a spacing of 600 ns per 1 macro pulse. The beam extracted from the RCS is delivered to the spallation neutron source through the 3 GeV RCS to Neutron Facility Beam Transport (3NBT) [2–4].

The linac comprises a volume-production type of H^- ion source, a 50-keV low energy beam transport (LEBT), a 3-MeV, 324-MHz Radio-Frequency Quadrupole (RFQ) linac, a 50-MeV, 324-MHz Drift-Tube Linac (DTL), a 200-MeV, 324-MHz Separated DTL (SDTL), and a 400-MeV, 972-MHz high-energy linac [5]. The RCS was designed to have a lattice with three-fold symmetry, resulting in three long straight sections [5]. One is dedicated to the long RF acceleration section, another to the injection and collimation, and the last to the extraction. In the RCS, 24 bending magnets, 60 quadrupole magnets and 18 sextqupole magnets were installed. The innovative development of the accelerating cavity loaded with magnetic alloy [6] enabled to solve the issue of powerful RF accelerating system.

For the pulsed spallation neutron source, liquid metal, mercury was employed as the target material because it has advantages in producing neutrons with its large atomic number of 80 and high weight density of 1.36×10^4 kg/m^3 via proton induced spallation reactions, and high heat removal capacity. Energies of spallation neutrons are in the MeV regime, hence they have to be slowed down to the cold neutron (meV) region suitable for neutron scattering experiments. For that purpose, liquid hydrogen (20 K) was selected as the moderator material. A mercury-target-moderator-reflector system was designed carefully in view of maximizing the intensity of neutrons and making the pulse shape as narrow as possible. As a result, 3 types of moderators, coupled, de-coupled and poisoned moderators were optimized to use 100% para-hydrogen.

Figure 1 shows a cutaway view of the spallation neutron source. The high radiation zone including the target, the three moderators in wing configuration, and the reflector and central shielding are placed inside a helium-filled vessel. The helium vessel with its inactive helium filling prevents the production of corrosive gases such as O_3, NO_x and acts as protective boundary against possible spills of mercury and light water in case of serious failure incidents. The moderators' side surfaces are the origins of the neutron beamlines. Since 2 sides of each moderator are used as viewed surfaces by the neutron beamlines, there are 6 origins, and 23 neutron beamlines are arranged from those origins. Outside the helium vessel, a neutron beam shutter made of steel with 2 m in length and 4 m height was installed on each neutron beamline. It moves vertically to open and/or close the beamline with a 1-m stroke.

The target, moderators and reflector are highly damaged by a high-power operation at 1 MW. Therefore, they must be replaced by full remote-controlled tools before the estimated end of lifetime. Since the lifetime of the target vessel is much shorter than that of the moderators and reflector, different access ways, horizontal access for the target vessel replacement and vertical access for the moderator-reflector system replacement, were selected. This enables us to replace the mercury target vessel without handling the moderators and reflector. The shielding structure surrounding the target-moderator-reflector system, e.g., neutron beam shutter, was determined by three-dimensional Monte-Carlo simulations in which geometry of all components was exactly modelled [7]. In particular, neutron and/or photon streaming through the gap between the components installed side by side were estimated carefully to meet the shielding design criteria that the dose rate was less than 2.5 μSv/h at the biological shield surface. As a result, the diameter of the target station was determined as 10 m with additional heavy weight concrete with a thickness from 2 to 2.5 m as biological shield, while the height was set to 10 m including an open space of 1.8 m for the neutron shutter drive devices.

Outside the target station, 3 GeV proton beam transport line components were installed at 330 m length upstream of the target. A cryogenic hydrogen system composed of a helium refrigerator and a hydrogen loop is arranged on the high bay from which liquid hydrogen is provided to three moderators. A target system maintenance area is placed just downstream from the target station where the components are replaceable with remote handling tools. The moderators and reflector are transferred into this are on the way of the high bay in case of the replacement. The maintenance scenario for the used components is summarized in reference [8]. Furthermore, primary and secondary cooling system and an off-gas processing system were installed in the areas next to the target maintenance area.

Construction of mega-watt class pulsed spallation neutron source is a challenging work. The Spallation Neutron Source (SNS) [9] project at Oak Ridge National Laboratory (ORNL, Oak Ridge, TN, USA) that was aimed at operating with a proton beam power of 1.4 MW at 60 Hz had been started a few years earlier than J-PARC. The European Spallation Source (ESS, Lund, Sweden) [10], a 2.86 ms long proton pulse facility designed to be operated with a 5 MW proton beam at 2 GeV, is under construction in Lund, Sweden. The concept to use a mercury target with liquid hydrogen moderators was already adopted in the SNS. However, details of our design are quite different from SNS. This article focuses on designs of the mercury target system, the moderators and the cryogenic hydrogen system are focused. They are described in the Sections 2–4, respectively. Furthermore, outline of the 3NBT is given in the Section 5.

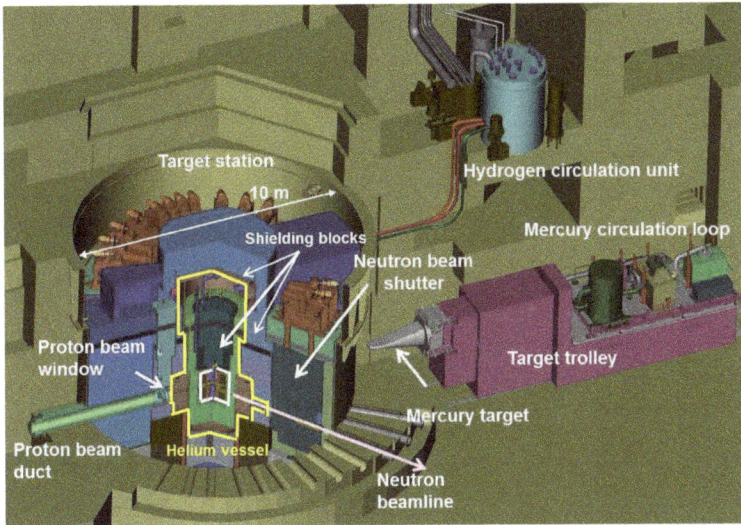

Figure 1. Cutaway view of the target station of pulsed spallation neutron source of the Japan Proton Accelerator Research Complex (J-PARC) with target trolley moved back to the service position.

2. Mercury Target System

2.1. Target Trolley

The target trolley is the carriage of the target vessel and the mercury circulation system as shown in Figure 2. It has a dimension of 12.2 m in length, 2.6 m in width and 4 m in height, and total weight of 315 tons and it installs the target vessel into the helium vessel for beam operation with enough radiation shielding behind against secondary particles generated via spallation reactions in mercury. For maintenance, it withdraws to the target maintenance area of hot cell (Maximum distance: 23 m). Radiation shield blocks made of iron and concrete provides most of the weight, through which many pipes such as mercury pipes and helium gas supply pipes penetrate. Behind the radiation shield, further iron blocks cover two mercury drain tanks and two spilt mercury tanks for shielding γ-rays emitted from radioactive spallation products in mercury in the drain tanks during the maintenance [1], and a trolley driving mechanism is mounted at the rear end. The mercury circulation system placed on the mercury system trolley comprises mercury pipings, a mercury pump, a heat exchanger, a surge tank, a gas supplying system and sensors.

In order to maximize the neutronic performance, the components are installed in the target station with minimum spatial gap. For example, the design gap between the neutron reflector housing and the target vessel was set to 8 mm considering the following assembling and positioning tolerances of those components; (1) the tolerance of the neutron reflector housing is 3 mm, (2) the manufacturing and assembling accuracy of the target vessel itself is 1 mm, (3) the positioning reproducibility of the target trolley is 1 mm, (4) positioning tolerance of the target vessel on the target trolley is 2 mm, and (5) an additional margin of 1 mm.

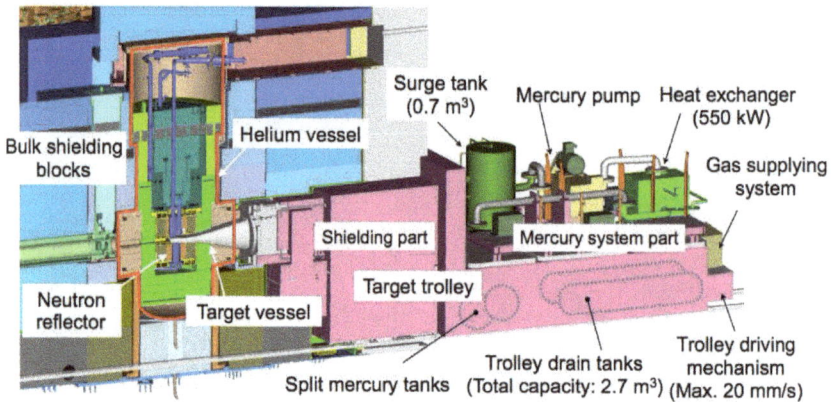

Figure 2. Layout of devices in the mercury target system mounted on the target trolley.

2.2. Mercury Target Vessel

Figure 3 shows photographs and schematic illustrations of the mercury target vessel. It comprises a mercury vessel and a double-walled water shroud covering the mercury vessel, which makes the target vessel a triple-walled structure. The forefront wall of the target vessel is called as the beam window on which proton beams are injected. The triple-walled structure prevents the mercury leakage to the outside in case of failure of the target beam window [11]. The mercury vessel is cooled by mercury flow, while the water shroud is cooled by water flowing through the double-wall gap. The interstitial space between the mercury vessel and the water shroud is filled with helium gas. In the mercury vessel, six flow vanes are placed in the mercury flow path to establish the "cross-flow", which indicates that mercury crosses the main part of the target in order to remove the heat generated by proton-induced spallation reactions in mercury as shown in Figure 3c. This cross-flow type (CFT) target vessel has definite advantages in terms of heat removal and mechanical strength. Mercury flow distribution in the CFT target, which corresponds to the heat generation distribution along the target length, enables efficient cooling of the spallation heat and small amount of mercury inventory. In addition, flow vanes are also designed as a reinforcement to ensure the enough mechanical strength of the vessel.

The inner structure arrangement of the target vessel was determined based on the results of thermal-hydraulic analyses, structural analyses and pressure wave analyses under the condition of 3 GeV, 1 MW proton beam injection. Abrupt temperature rise in mercury caused by the pulsed proton beam injection generates pressure waves. The pressure wave load and cavitation erosion induced on the beam window has been recognized as the critical issue for the target vessel design [12]. The original target vessel was designed on the condition that the proton beam profile is Gaussian distribution with a beam footprint of 130 mm × 50 mm. Recently the proton beam profile was changed to the beam footprint of 180 mm × 70 mm to moderate the pressure wave load on the beam window, decreasing the maximum heat generation rate down to 220 MW/m^3 from the former value of 668 MW/m^3. Furthermore, in order to increase the durability of the beam window against the cavitation erosion, the inner surface of the beam window was hardened by using carburizing method named Kolsterising®, Bodycote plc, Macclesfield, UK [13]. The bubble generator to inject gas micro-bubbles in mercury is installed in the target vessel to reduce the pressure waves. Details will be described in Section 2.3. Furthermore, a double-walled structure target with a narrow gap mercury flow channel of 2 mm in height at the beam window was adopted since 2013, aiming at mitigating the cavitation damage formation by increasing the mercury flow velocity and the narrow gap effect [14].

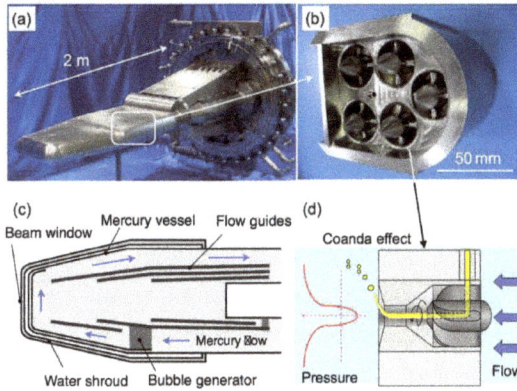

Figure 3. Photographs of target vessel (**a**) and bubble generator assembly (**b**), and their schematic views (**c**) and (**d**).

2.3. Microbubble Injection System

Currently, cavitation damage is considered to be the dominant factor to determine the service lifetime of the target vessel rather than radiation damage. Non-condensable helium gas micro-bubbles are effective to suppress the pressure waves in mercury which causes cavitation, because they absorb thermal expansion of mercury at the proton beam injection, and change kinetic energy of the pressure wave to thermal energy by their oscillation. Since pressure rising is very fast, e.g., the maximum pressure reaches to 40 MPa at 1 μs after the 1 MW proton beam injection, it is necessary to inject gas micro-bubbles less than 100 μm in diameter with a 0.1% volume fraction to mercury for effective pressure waves mitigation [15,16].

We developed a gas microbubble generator [17] for generating bubbles to satisfy the design condition mentioned above, and have installed it in the mercury target system with a closed-loop gas supply system in October 2012. Figure 3b,d show the photograph and schematic of the microbubble generator. Gas is injected from the center of the static swirler to make a gas column and brake down to the microbubbles owing to the shear force induced by the vortex-breakdown at the outlet of bubble generator. Multiple bubble generators with opposite swirl direction were placed alternately to prevent the bubble coalescence due to the bulk swirl flow.

The gas supplying system circulates the helium gas enclosed in the mercury loop as follows: helium gas flows from upper space of the surge tank to a compressor of gas supplying system and is pressurized, so that it could flow towards the bubble generator in the target vessel according to the differential pressure between those components. Double containment metal bellows compressor was selected for the gas supplying system to detect the gas leakage from internal bellows and to assure the containment of radioactive cover gas in it. The flow rate of helium gas was adjusted to be 1.5 L/min at standard condition with a flow control valve. Resultant helium gas fraction at the beam window of the target vessel was estimated to be about 1.5×10^{-4} under this flow condition.

2.4. Mercury Circulation System

As mentioned before, primary components of the mercury circulation system are the surge tank, the mercury pump, and the heat exchanger. The most important and challenging issue for this system design was that all the components should be maintained by remote handling. Though the system seems very simple in the figure, it is actually more complicated because a lot of devices such as valves, thin pipes, cables, flanges, connectors etc. are installed to the system, which are not shown in Figure 2. Thus, each component was devised and arranged on the target trolley, and so that it could be maintained easily by remote-handling. Table 1 summarizes primary specifications of each component.

The mercury circulation system will be operated with a mercury pressure of 0.5 MPa, with keeping this operational pressure by pressurizing the surge tank with helium gas. Type 316L stainless steel (316L SS) was chosen as the primary component material considering its corrosion and radiation resistant property. Inner diameter of the mercury primary piping is 143.2 mm and no valves are installed on the primary loop. Also, the mercury loop has neither strainer nor filter, because the surge tank is expected to serve the same function. Most of mercury compounds produced with spallation product or component of wall material in the system are solid and lighter than mercury itself, so they will be trapped in the surge tank as scum floating on mercury, which was demonstrated in a small mercury loop in our laboratory. In order to minimize the mercury piping erosion and prevent the mercury cavitation, the maximum mercury flow velocity was set to be less than 1 m/s. As a result, the operational mercury flow rate was determined as 41 m^3/h, which was equivalent to the mercury flow velocity of 0.7 m/s in the mercury primary piping. Based on the experimental data of erosion and corrosion by mercury flow, pipe wall thickness of 11 mm was chosen to have enough erosion and corrosion allowance for the designed facility life time of 30 years [18].

For circulating mercury, a permanent magnet rotating type induction pump (PM pump) is employed. The PM pump can be operated at a low risk springing a mercury leak because it has no seal parts. However, a large motor is required to generate sufficient flow rate on account of the low efficiency of the PM pump. Almost the total motor power is converted to the thermal heating in the PM pump, which should be removed by a heat exchanger in the mercury circulation system. Therefore, the PM pump was developed to reduce the thermal heating and to increase the flow rate of the mercury [19]. Figure 4 shows a schematic drawing of the PM pump. The dimensions of the pump are 0.8 m in length, 1.1 m in width, and 2.0 m in height, and total weight is 2.0 tons. In this pump, the mercury duct was made in a circular shape and rotating permanent 16 poles Sm-Co magnets were installed at the center of the duct to apply the Lorentz force, and the motor to rotate the permanent magnets was vertically set on the top of the pump to arrive at a compact PM pump design. Furthermore, the thickness of the duct wall and the duct width of the practical PM pump were carefully determined to assure the pump pressure to circulate mercury in required flow rate taking account of a power loss from the 90 kW motor for the thermal heating in the duct wall. The resultant duct is made from 316L SS with a diameter of 379 mm, a height of 340 mm, and an inner and outer thickness of 3 mm and 5 mm, respectively.

A double walled plate type heat exchanger of which parts are fully welded and fabricated together was adopted in order to reduce the possibility of radioactive mercury leak to the secondary loop. Helium gas is sealed in the double wall gap and mercury leak into the gap will be detected by monitoring the helium pressure change. The mercury is cooled down from 72 °C to 50 °C in this heat exchanger. The secondary coolant is light water with an inlet temperature of 35 °C and it is cooled in the cooling tower placed outside of the facility building.

Table 1. Primary specifications of mercury circulation system components.

Name of Component	Specification
Mercury Pipe	Inner Diameter: 143.2 mm
	Wall thickness: 11 mm
Surge tank	Capacity: 0.7 m^3
Mercury pump	Type: Permanent magnetic type
	Rated flow rate: 41 m^3/h
	Discharge pressure: 0.5 MPa
Heat exchanger	Type: Double wall (All welded)
	Heat removal capacity: 600 kW (max.)
Flow Meter	Type: Venturi
	Measurement range: 0~60 m^3/h

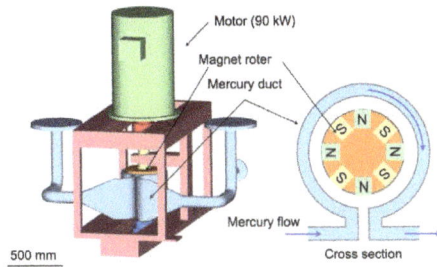

Figure 4. Schematic view of the permanent magnet rotating type induction pump (PM-pump).

2.5. Target Diagnostic System

In order to monitor the pressure wave-induced vibration of the target vessel at the proton beam injection, a non-contact and nondestructive diagnostic method is required to be used in the high-radiation environment [20]. To meet the requirement, a novel in situ diagnostic system using a laser Doppler vibrometer (LDV) was developed. Figure 5 shows the schematic drawing of the LDV diagnostic system. It comprises a retro-reflecting corner-cube mirror (reflective mirror) mounted on the upper surface of the target vessel, a mirror assembly, a laser light source and an avalanche photo diode as a laser light detector [21]. Laser light is emitted on the top of shielding plug placed above the target vessel and passes through the mirror assembly for 3.4 m and incident upon the reflective mirror on the target vessel. A He-Ne laser (wave-length: 632 nm) with a power of 2 mW was employed to have high coherence to improve the signal to noise ratio (S/N). The reflective mirror was fabricated out of a Au plate with a direct micro machining process to have high-reflectance of 56% and high-corrosion resistance.

Figure 6 shows the displacement velocity of the target vessel measured with LDV for the 310 kW proton beam injection with a power density of 2.8 J/cm^3/pulse [22]. It is obvious that the maximum displacement velocity observed after injecting proton beam decreased about one third by injecting gas micro-bubbles, and also decreased significantly in the time region after 2 ms.

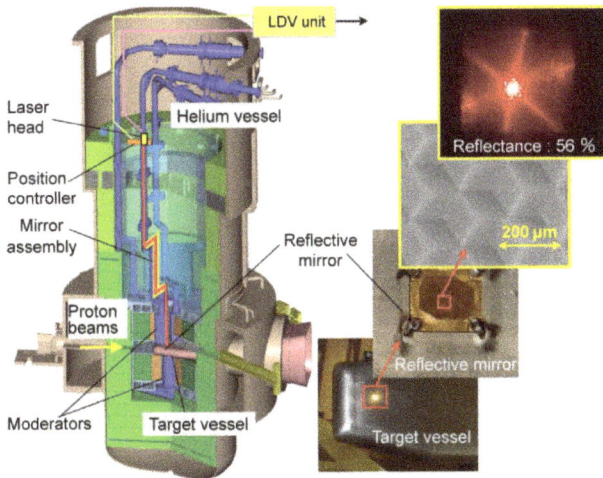

Figure 5. Target diagnostic system consists of the Laser Doppler vibrometer (LDV) installed in the helium vessel.

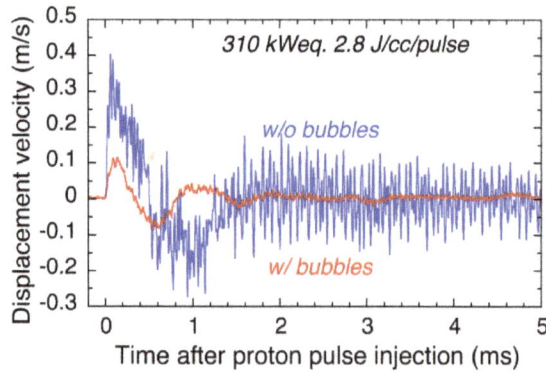

Figure 6. Time histories of the displacement velocity of the mercury target vessel measured with LDV system. Time 0 designates the proton impact on the target. Blue and red lines represent the data without and with injecting gas microbubbles, respectively.

2.6. Target Vessel Replacement

Once the target trolley is withdrawn into the maintenance room, the component replacement or repairing work is carried out using a power manipulator and some master-slave manipulators. Before removing the target, cover gas in the surge tank was transferred to an off-gas process system to reduce the radioactivity of gaseous radioactivity, especially ^{127}Xe and ^{3}H, accumulated in the cover gas. The transferred cover gas is stored in gas holders of the off-gas process system for about 12 months waiting for the decay of ^{127}Xe having half-life of 36.4 days, while ^{3}H is absorbed by a molecular sieve (synthetic zeolite) [23], and then released to the environment after checking the radioactivity. This gas transfer operation is repeated until the radioactivity of ^{127}Xe decreased down to 1/100 of its initial value, which is significantly below the allowable value even if the remaining ^{127}Xe in the loop is released through the stack of the building at any failure.

Figure 7 shows the procedure of the mercury target vessel replacement. A storage container is mounted on a target exchange truck. It supports the mercury target vessel after the connecters are disconnected and bolts are loosened. It is noted that mercury was drained into the drain tank installed beneath the target trolley prior to dismounting the target vessel. The used mercury target vessel in the storage container is moved to a storage room located below ground. The new mercury target vessel is installed to the target trolley in the reverse order of the target vessel removal. After the installation, airtightness test on all pipe connections is conducted to ensure their enclosure performance by filling helium into the circulation loop, where the criterion of airtightness is 10^{-6} Pa m^{3}/s. The helium filled into the loop is also transferred to the off-gas process system and then mercury is filled into the circulation loop from the drain tank.

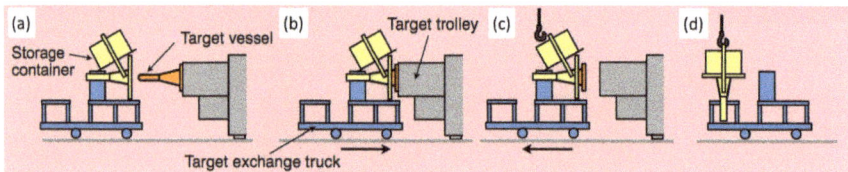

Figure 7. Procedure of the mercury target vessel replacement. (**a**) preparing the target exchange truck and the storage container; (**b**) insert the target vessel into the storage cask; (**c**) removing the target vessel from target trolley; (**d**) move the storage cask to the storage room. New target is set to the target trolley in the reverse order.

3. Moderator and Reflector Development

The wing geometry type, as shown in Figures 8 and 9, was adopted for the moderator-target arrangement where the horizontal proton beam is injected on the target. It was widely utilized at pulsed neutron sources, such as KENS facilities of High Energy Accelerator Research Organization (Tsukuba, Japan), ISIS of Rutherford Appleton Laboratory and the SNS of ORNL. At J-PARC, three moderators are arranged in total, two decoupled moderators (decoupled and poisoned) above the target and one coupled moderator below the target, respectively. The coupled moderator provides high peak intensity in neutron beam pulses by setting it on the target vessel closely without thermal neutron absorber. The decoupled moderator provides neutron pulses with short tails by adopting a thermal-neutron-absorber-material called decoupler although the neutron intensity is sacrificed somewhat with it. For poisoned moderator, a sheet made from thermal-neutron-absorber is installed inside the moderator vessel so that the neutron beam pulse could have both narrow width and short tail. Illustrations of the coupled moderator and the decoupled moderator are shown in Figures 10 and 11. The moderator specifications are listed in Table 2. The arrangement (upper or below) was chosen to separate the coupled moderator from the thermal neutron absorber area installed for two decoupled moderators, and to provide the neutron scattering instrument, reflectometer, with neutrons through a vertically inclined neutron beam line that is optimized to use the coupled moderator.

Resultant three moderators have following distinctive features: moderators were optimized to use para-hydrogen, giving highest neutronic performance, such as neutron intensity per pulse, especially for the coupled moderator,

1. high decoupling energy of 1 eV was achieved by adopting a Ag-In-Cd decoupler, providing short tail in the pulsed neutron beam emitted from the decoupled and poisoned moderators,
2. cut-off energy was extended up to 0.4 eV by adopting a cadmium poison sheet, resulting in sharper pulse width for the poisoned moderator.

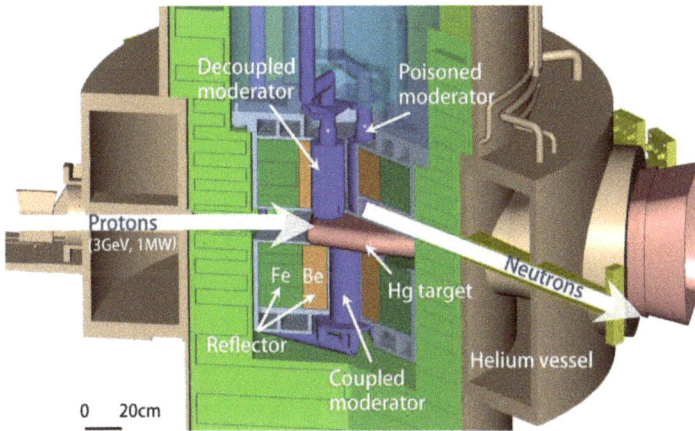

Figure 8. Cross sectional view of target-moderator-reflector.

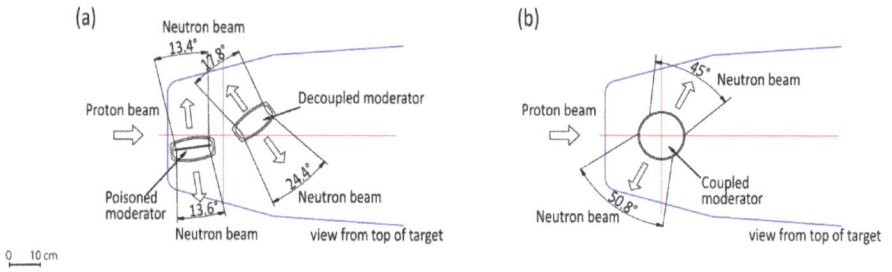

Figure 9. (**a**) Locations of decoupled and poisoned moderators with respect to the target vessel; (**b**) Location of coupled moderator with respect to the target vessel. The directions and angles of neutrons extracted from each moderator surface to neutron beamlines are also shown.

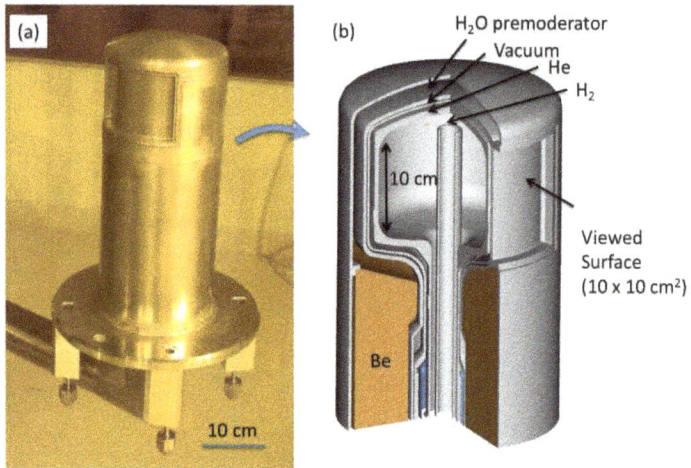

Figure 10. (**a**) Photograph of coupled moderator; (**b**) Cutaway view of coupled moderator.

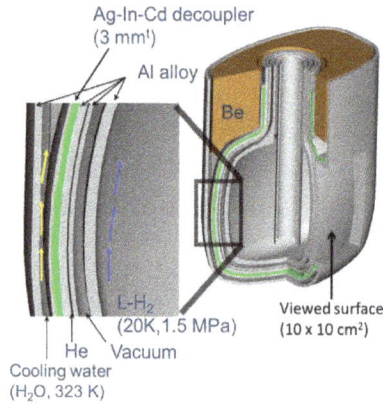

Figure 11. Illustration of inside structure of de-coupled moderator.

Table 2. Major specifications of moderators employed at J-PARC.

Moderator	Coupled	Decoupled	Poisoned
Beam extraction	Two-sided	Two-sided	Two-sided
Angular coverage	50.8°/45°	24.4°/17.8°	13.4°/13.6°
Main moderatow	H_2, 20 K, 15 atm	H_2, 20 K, 15 atm	H_2, 20 K, 15 atm
Shape	Cylindrical	Canteen	Canteen
Size	$\phi14$ cm $\times 10^H$ cm	$13^W \times 12^H \times 6.2^t$ cm^3	$13^W \times 12^H \times 6.2^t$ cm^3
Premoderator	H_2O (1 cm^t)	Non	Non
Decoupler	-	Ag-In-Cd	Ag-In-Cd
Decoupling energy		1 eV	1 eV
Poison	-	-	Cd (1.3 mm^t)
Viewed surface	10×10 cm^2	10×10 cm^2	10×10 cm^2

H: Height; W:Width; t:thickness.

To obtain an optimum for the target-moderator-reflector layout in view of enhancing neutron production efficiency, design parameters such as the proton beam footprint on target, relative longitudinal moderator position to the target and distance between target and moderator were assessed by neutronics calculations. As a result, the proton beam footprint with 13×5 cm^2 at the target front was determined. It was also estimated that intensity of the slow neutron emitted from the coupled moderator was maximum at 15.5 cm downstream from the target front [1,24,25] which determined the coupled moderator position with respect to the target. The distance between target and moderator were set to 8 mm taking account of tolerances in component fabrication and installation as mentioned in Section 2.1.

Liquid hydrogen has been identified as the only practical cold moderator material for the moderation of MW-class high-intensity sources, considering the irradiation damage [26–28]. Since it has a relatively low hydrogen density (ca. on half of solid methane), it has not been used widely in view of enhancing cold neutron intensity in the spallation neutron source so far. In this design, we focused on the para-hydrogen, one of two isometric forms of liquid hydrogen, to improve the neutronic performance, because it would give shorter tails on the resulting neutron pulses due to its low cross section blow 14.5 meV [29]. On the other hand, it was estimated that neutrons leaked more quickly in the slowing down process. To overcome the drawback of hydrogen for an MW-class energy source, extensive optimization studies have been conducted [30–40]. Especially for the coupled moderator, unique optimization to have large sized cylindrical shape coupled with optimized water pre-moderator gave higher neutron intensity [36,37], and so could provide the highest neutron intensity per pulse in the world [40]. It was also concluded that para-hydrogen fraction plays an important role of neutronic performance [37,40]. This benefit of para-hydrogen moderator is taken into account in the present high power neutron source projects of the ESS [41,42] and the second Target station design at SNS of ORNL [43].

As mentioned above, the neutronic performance, particularly pulse shape, strongly depends on the para-hydrogen fraction [35,38], as shown in Figure 12. E. B. Iverson and J. M. Carpenter calculated the irradiation effect on the para-hydrogen to ortho-hydrogen conversion under the high neutron irradiation environment [44,45]. Changes in the pulse shape of the para-hydrogen fraction were measured at the neutron source at Hokkaido University Electron Linac Facility and the Manuel Lujan Jr. Neutron Scattering Center of the Los Alamos National Laboratory [46,47]. However, there was no clear indication of the irradiation effect on para-hydrogen to ortho-hydrogen conversion because the neutron intensity was too weak at the Hokkaido University facility and the pulse shapes emitting from the moderator were only measured according to the elapsed proton beam operation time in the Los Alamos experiment. Then, we developed a method of sampling gaseous hydrogen from the circulating hydrogen in the hydrogen loop (1.5 MPa, 20 K) under high neutron irradiation at J-PARC, and measure the para-hydrogen fraction with a Laser Raman spectroscopy. The measured para-hydrogen fraction was consistent with the equilibrium level of para-hydrogen at 20 K (99.8%) at the 300 kW beam operation because a $Fe(OH)_3$ catalyst worked well in the loop. The measured pulse shape at 300 kW was also in good agreement with the calculated result for the 99.8% para-hydrogen [48].

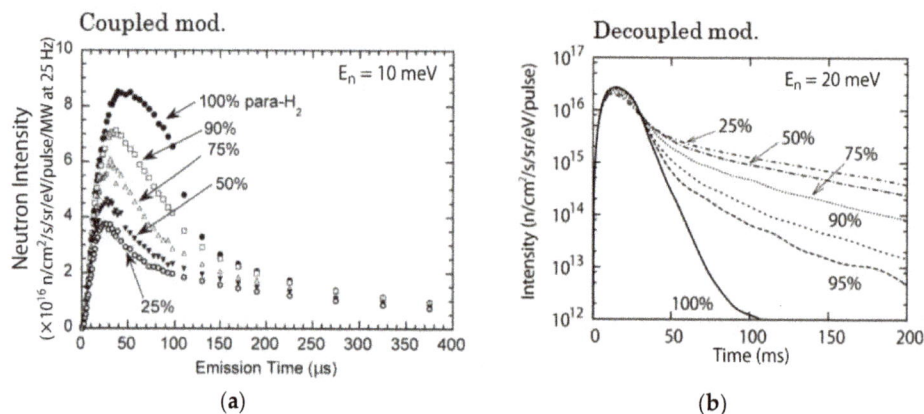

Figure 12. (a) Para-hydrogen concentration dependence on pulse shapes for the coupled moderator [30] and (b) the de-coupled one [38].

For the decoupled and poisoned moderators, a thermal neutron absorber acting as decoupler as shown in Figure 11, is installed around the moderator except for neutron beam extraction window, resulting in a short decay time in neutron beam pulse. Higher cut-off energy, which is called decoupling energy (E_d), gives a shorter decay time [49]. At J-PARC, neutron beam users required at least 1 eV decoupling energy for high-resolution neutron scattering experiments in the design stage. At the ISIS of Rutherford Appleton Laboratory (160 kW neutron source), boron-based material, such as sintered boron carbide (B_4C), was previously utilized to obtain a high decoupling energy (over 1 eV) with controlling its thickness. However, this cannot be used for MW-class sources because of the helium embitterment caused by the (n, α) reaction. In order to realize a high decoupling energy of 1 eV, we focused on a Ag–In–Cd alloy [50], a combing elements of different resonance energy absorption to yield an effective decoupling energy of 1 eV. Ag-In-Cd sheathed with stainless steel is already utilized for the control rod of PWR (pressurized water reactor). In view of heat removal and corrosion protection, however, there was the issue that the Ag-In-Cd plate has to be bonded to the Al alloy (A5083), the structural material of a moderator and reflector. We succeeded to implement Ag-In-Cd alloy into the moderator-reflector as a decoupler by adopting HIP (Hot Isostatic Pressing) method [49,51,52]. In order to validate neutronic performance of the Ag-In-Cd decoupler, the pulse shape was observed in the rise and tail part of the Bragg peaks. As shown in Figure 13, the measured and calculated time structures of the neutron pulses were in good agreement [40]. However, the Ag–In–Cd decoupler has a disadvantage that highly residual radioactivity is generated especially by the production of ^{108m}Ag (half-life of 418 years).

We proposed a new idea [53] to reduce the potential risk of high residual radioactivity of AIC. It was that Au replaced silver in the Ag-In-Cd composition. As a result of neutronic calculation considering the component change accompanied with accumulation and decay of residual nuclides, Au-In-Cd could reduce the residual radioactivity by three orders of magnitude than Ag-In-Cd without sacrificing neutronic performance [53]. We also adopted HIPing method to implement the Au-In-Cd decoupler to the next moderator-reflector fabrication [54–56]. The next moderators with the Au-In-Cd decoupler are under fabrication.

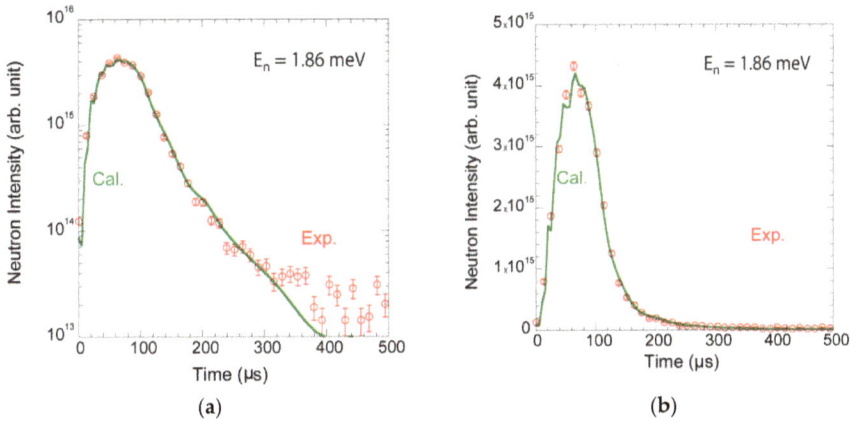

Figure 13. Measured and calculated neutron pulse emitted with 1.86 meV from decoupled moderator [40]. Graph is shown by logarithmic scale (**a**) and linear scale (**b**) in vertical axis. Open circle and solid line represents measured data and calculated results, respectively.

For the poisoned moderator, a Gd sheet has been widely used as a poison material. It was adopted at the Intense Pulsed Neutron Source (IPNS) in Argonne National Laboratory (Chicago, IL, USA), ISIS, and KENS, and is also in use at SNS. However, we attempted to use a Cd poison [39] thinking of two advantages. One is a higher cut-off energy than Gd, leading to make a sharper pulse width at higher neutron energies up to 0.4 eV, and the other is a weaker resonance capture leading to make the intensity penalty with thicker poison plate much smaller. By optimizing the Cd based poisoned moderator, actually outstanding performances of the world's highest resolution [40], 0.035% in $\Delta d/d$ was obtained at the super high-resolution powder diffractometer (SuperHRPD) [57] with long neutron flight path of 94.2 m, where d is the lattice-spacing of the Si-400 reflection at 4.628 Å^{-1}. The diffraction pattern obtained at the SuperHRPD is shown in Figure 14.

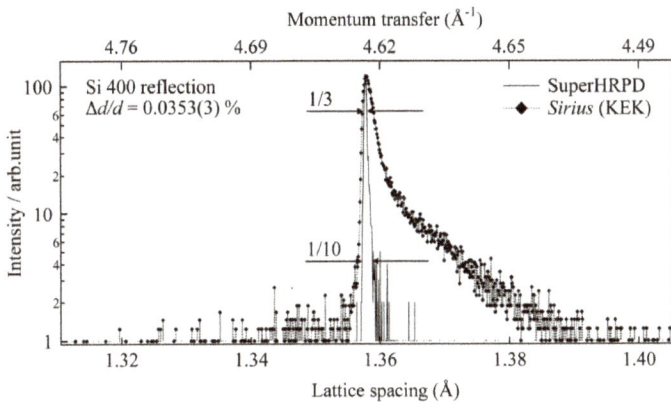

Figure 14. Comparison of neutron diffraction data for a silicon sample measured at super high-resolution powder diffractometer (SuperHRPD) (solid line) with those obtained at Sirius (dots) [40].

4. Cryogenic Hydrogen System

The cryogenic hydrogen system plays the role of providing the three moderators with cryogenic hydrogen continuously. It was estimated that several kW of heat was generated in the moderators through nuclear heating at the 1 MW operation. This required the cryogenic system to have a kW-power. Forced-flow circulation cooling is also adopted to maintain the temperature of the moderators at 20 K regime and to attain small temperature difference between the inlet and the outlet of the moderator. The design of the system configuration, the specification of the components, and the physical layout are introduced as follows [58].

4.1. Design Requirements

The hydrogen moderator makes fast neutrons generated at the mercury target slow down to cold neutrons. The cryogenic system was designed to have following performance:

(1) Average temperature at the moderators is kept less than 20 K at 1.5 MPa in order to achieve constant density.
(2) Temperature difference between the inlet and the outlet of moderators is kept less than 3 K to prevent hydrogen density fluctuation.
(3) Para-hydrogen concentration of the moderators is kept more than 99% for the neutronic performance that helps to slow down to cold neutrons.

From the viewpoints of safety, the following additional requirements were taken into account:

(4) In order to reduce hazard potentials, it is necessary to minimize the inventory of hydrogen in the system.
(5) Blanket structure with inert gases needs to prevent the hydrogen leakage to the outside from the system and to prevent air from entering the system.
(6) Hydrogen vent line is required to release hydrogen safely when the refrigerator operation stops and any off-normal event takes place.

To meet the requirement of the item (4), we designed the cryogenic system comprising a hydrogen circulation system with the moderators and a helium refrigerator to cool it, because the helium refrigerator had been widely used in recent years. In order to satisfy the operational temperature condition (1), we decided to flow the cryogenic hydrogen in parallel to the three moderators. Since the hydrogen circulation system is a closed loop as a refrigerator, a pressure control system is required to compensate a large pressure change. If there is no control of pressure, the pressure change of about 3.7 MPa takes place at the moment that the accelerator beam is turned on and/or off at the 1 MW operation. Table 3 shows the nuclear heating power in the moderators and their vessels estimated by the neutronic design study. Based on these conditions and requirements, process flow of the cryogenic hydrogen system was calculated by solving mass and heat balance at the components to determine the sizing and the specifications. It is noted that we have to design, operate and maintain this system according to the Japanese regulation, High Pressure Gas Safety Lows.

Table 3. Estimated nuclear heating power deposited on each moderator at 1 MW proton beam.

	Coupled Moderator	Decoupled Moderator	Poisoned Moderator	Total
Volume (L)	1.54	0.97	0.97	3.84
Nuclear heating in hydrogen (W)	946	467	442	1855
Nuclear heating in vessel (W)	793	519	584	1896
Total heating in the moderator (W)	1739	986	1026	3751

4.2. Design of the Cryogenic Hydrogen System

4.2.1. System Configuration and Specification of the Main Components

A schematic configuration of the cryogenic hydrogen system is shown in Figure 15. It comprises a hydrogen circulation system to cool high-energy neutrons at the moderators and a helium refrigerator system to cool hydrogen in the hydrogen circulation system. Figure 16 shows the arrangement of the cryogenic hydrogen system in the Materials and Life Science Experimental Facility (MLF) building. The compressor and related equipment, the cylinder storage for gas supply are placed outside the building. The cold box is arranged as close as possible (within 10 m) to the hydrogen circulation system to minimize the heat loss. The moderators are located at the center of the neutron source station and connected with the hydrogen circulation system with hydrogen transfer tubes of about 14 m length. They are able to be removed from the transfer tube with a coupler on top of the helium vessel in case of the replacement.

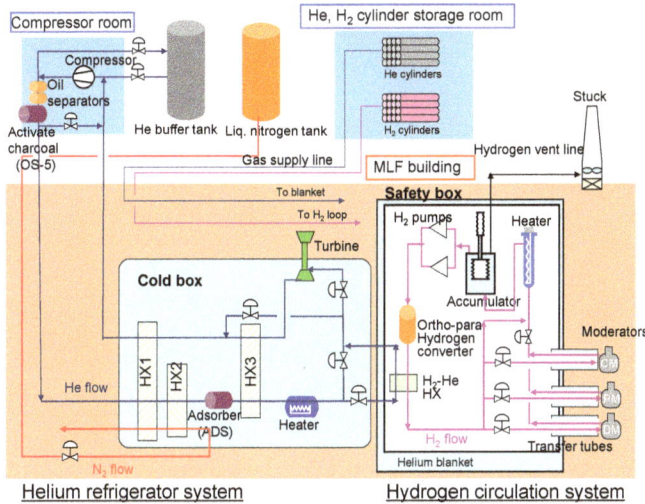

Figure 15. Schematic configuration of the cryogenic hydrogen system employed for pulsed neutron source of the Japan Proton Accelerator Research Complex (J-PARC).

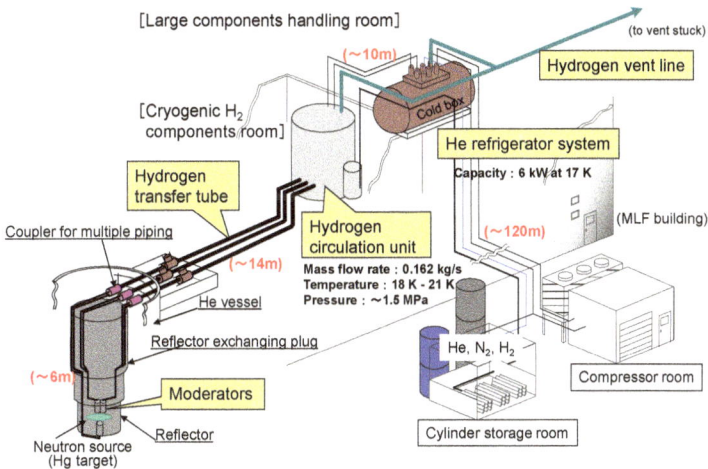

Figure 16. Illustration of cryogenic system component arrangement in Materials and Life Science Experimental Facility (MLF) building.

4.2.2. Helium Refrigerator System

The helium refrigerator system comprises a helium compressor, oil separators including an activated charcoal as an oil adsorber, a helium buffer tank and a cold box containing three heat exchangers, an adsorber, heater and turbine in the vacuum vessel. Table 4 summarizes specifications of the main components. They were determined based upon the results of the process flow calculation for the helium refrigerator system. The helium-refrigerator's refrigerating power at 17 K is specified to be around 6 kW in which a margin of 17% for the estimated total heat load is considered. In the helium refrigerator, an expansion turbine is located after the He-H_2 heat exchanger so as to keep hydrogen pressure to be 1.6 MPa in the helium part that is higher than the pressure of less than 1.5 MPa in the hydrogen part, enabling to prevent hydrogen leak into the refrigerator. Moreover, a heater installed prior to the He-H_2 heat exchanger keeps the hydrogen outlet temperature of the heat exchanger in constant.

4.2.3. Hydrogen Circulation System

The hydrogen circulation system circulates the cryogenic hydrogen through the three moderators and removes the heat generated in the moderator. It is made up of a He-H_2 heat exchanger, two hydrogen pumps, an Ortho-Para hydrogen converter, an accumulator and a heater. Those components are contained in a vacuum vessel, called safety box, covered by a helium blanket. Table 4 summarizes specifications of those components. The absorbed heat is transferred to the helium refrigerator through the He-H_2 heat exchanger.

The hydrogen pump is designed to get the mass flow rate of 0.162 kg/s with a pump head of 0.12 MPa at the rated condition [59]. It is an essential component to ensure reliable operation. Two pumps with the same specification are arranged in parallel and operated simultaneously with 50% of the load, respectively, and so have redundancy to continue operation with single pump with a 100% load even if either pump has failed.

The Ortho-Para hydrogen converter plays the role of keeping para-hydrogen concentration at more than 99% in the moderators in terms of providing superior neutronic performance. It was designed that the about 2% of ortho-hydrogen, which was converted from the para-hydrogen by the neutron irradiation in the moderator during the operation, is immediately converted to para-hydrogen through the converter. Our cryogenic system is the first device to install the Ortho-Para hydrogen converter among MW-class spallation neutron sources in the world.

A heater and an accumulator are installed as the pressure control devise to compensate large pressure fluctuations generated in the closed loop of the hydrogen circulation system. The heater performs an active control for thermal compensation, and the accumulator plays a role of passive volume control [60]. This is also the first time to use them together in the cryogenic system in the world. Present pressure control device enables the heater power to be smaller than the case with only the heater, and the hydrogen inventory to be smaller than the case controlled with only the accumulator.

4.2.4. Hydrogen Vent System

The hydrogen vent system is a specific apparatus. Inert gas, such as helium or nitrogen, always purges inside of the vent line to avoid air ingress into it. When the operation of this refrigerator is completed or stopped by an off-normal event, entire hydrogen in the system is safely released to the outside with inert gas.

Table 4. Specifications for main components of the cryogenic system.

Helium Refrigerator System	
Cold box	Type: Helium Brayton cycle Refrigeration capacity: 6000 W at 17 K Supply pressure: 1.6 MPa Liquid nitrogen consumption: 103 L/h
Helium compressor	Type: Oil injection screw compressor Suction/Discharge pressure: 0.31/1.7 MPa Mass flow rate: 0.285 kg/s Motor ratings: 690 kW
Hydrogen Circulation System	
Hydrogen circulation pump	Type: Centrifugal pump Inlet temperature: 19 K (available for 300 to 17 K) Inlet pressure: 1.5 MPa (available for 0.5 to 1.8 MPa) Mass flow rate: 0.162 kg/s Pressure head: 0.12 MPa (at 0.162 kg/s) Revolution: 30,000 rpm to 60,000 rpm
Ortho-Para hydrogen convertor	Catalyze: Iron hydroxide ($Fe(OH)_3$) Outlet Ortho-Para concentration: 99.0% Catalyst filling volume: 35 liter
Pressure control system Accumulator Heater	Type: Bellows structure (Inner welding bellows, Outer molding bellows) Size: ϕ310/350 mm (Inner bellows), ϕ59/80 mm (Outer bellows) Design pressure: 2.1 MPa Variable volume: 3.5 L Type: Sheath heater with baffle plate Power: Max. 7 kW
Heat exchnger	Type: Aluminum plate fin Size: W200 × H120 × L700 mm Temperature: H_2 22 (in)/18 (out) K, He 17 (in)/20 (out) K

5. Proton Beam Transport

5.1. Overview of the Proton Beam Transport

Figure 17 shows a plan view from the 3 GeV Rapid Cycling Synchrotron to the MLF building. In the MLF building, the Muon Science facility (MUSE) [61,62] is arranged upstream of the spallation neutron source, where a 2-cm-thick carbon graphite target is used for the muon production at 33 m upstream of the neutron production mercury target in a cascade scheme. The proton beam loses about 6.5% by the nuclear interactions at the graphite target. The spread angle of the 3 GeV proton beam scattered at the graphite target is estimated as 1.25 mrad. Considering this beam characteristics, the beam optics from the muon production target to the neutron production target was designed that the beam loss could be suppressed less than 1 W/m at the 1 MW operation. It is observed further beam loss of about 1.5% between the muon production and the neutron production targets. The present cascade target scheme is thought to give an advantage over a separated target scheme, because time sharing of the delivered proton beam to the neutron and muon targets would provide only 50% of the source strength to the respective user community. The 3NBT has been operated for eight years since it delivered the very first proton beam pulse to the neutron source in 30 May 2008. In 2015, for the first time, single bunches of the 1-MW-equivalent proton beam pulse were delivered to the neutron production mercury target without significant beam loss in the accelerator study conducted in January 2015. Following this study, in April 2015, proton beam power for the user program was set to 500 kW at 25 Hz operation.

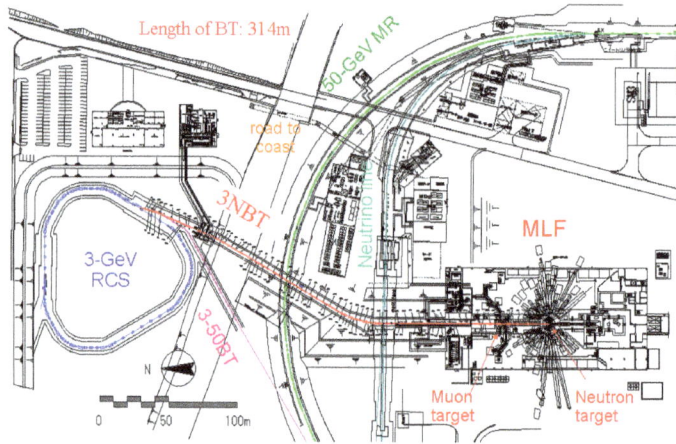

Figure 17. Plan view of the 3 GeV Rapid Cycling Synchrotron (RCS), the 3 GeV RCS to Neutron facility Beam Transport (3NBT) and the MLF at J-PARC.

5.2. Instruments for Beam Transport to Target

5.2.1. Magnets for Beam Transport

Since an uncontrolled high-power proton beam easily activates the beam transport equipment, a small beam loss rate is required in view of the instruments maintenance. In J-PARC, criterion for the beam loss was set as 1 W/m considering handling the instruments by hands on access. In order to deliver high intensity beam with very small beam loss, a careful design of the magnet, high precision and stable power supplies are essential. Also, it is important to align beam line components with high precision. Therefore, we routinely have been performing geometrical survey and alignment with a high precision laser tracker. Table 5 shows the specification of the magnets utilized for the 3NBT. Since high radiation environment is generated by the beam loss during the beam transport, it was required for the magnets to have high radiation resistance. Especially at the muon production target, large beam loss of about 6.5% is expected because of the nuclear interaction of proton beam with target nuclei. The insulator for the magnets installed around the muon target was carefully designed because an ordinary insulator such as polyimide used for coils will lose its insulation performance in the radiation environment. In the end, mineral insulator cable (MIC) made of MgO was utilized considering its high radiation resistance [63], enabling the absorbed radiation dose larger than the acceptable limit for polyimide of 400 MGy. It is also noted that the magnets around the muon production target can be replaced by remote handling if necessary in case of failure. They could be re-aligned on the exact position with an alignment pin installed at the bottom of the magnet.

Table 5. Specifications of magnet used in beam transport for spallation neutron source.

Magnet Type	Dipole for Bending	Quadrupole		Dipole for Sterling		Octupole
Insulator	Polyimide	Polyimide	MIC	Polyimide	MIC	Polyimide
Installed numbers	9	51	3	45	2	2
Aperture (mm)	160	220~300	260	200~300	260	300
Field or gradient	1.1~1.5 T	6~8 T/m	8 T/m	0.06 T	0.08 T	$800 \, T/m^3$
Field uniformity	5×10^{-4}	3×10^{-3}	3×10^{-4}	2×10^{-2}	2×10^{-2}	
and region (mm)	100	102~140	123	100	100	
Total weigh (t)	14~20	5~38 *	38 *	0.5~56 *	56 *	6

* Including shielding plug around muon production target. MIC: mineral insulator cable.

5.2.2. Proton Beam Monitor

In order to measure the characteristics of the proton beam, a diagnostic system with a Multi Wire Profile Monitor (MWPM), shown in Figure 18, was developed. Principle of the MWPM is simple that it measures the amount of electrons emitted by the interaction of the beam at the wire [64]. As a sensitive wire material, tungsten wire is generally selected because it emits a large amount of electrons and its melting point is so high. In the present system, however, SiC was chosen considering its high radiation resistance that can stand for the radiation damage up to 80 displacement per atom (DPA). Considering the high-intensity proton beam, proton beam loss at the MWPM is another important factor in selecting the material suitable for the wire. Since the Rutherford scattering of protons by target atom is proportional to the square of the atomic number, material with a low atomic number, such as SiC, has an advantage to reduce the beam loss.

Along the beam transport line, 15 sets of movable MWPMs were placed to measure the beam profile. The MWPM frame installed 31 wires of SiC with a spacing pitch of 6 mm in horizontal and vertical direction, respectively, where the SiC wire has a tungsten core of 0.01 mm covered with 0.15 mm of SiC. The wire frame was made of aluminum oxide with purity more than 95% to have enough high radiation resistance. The MWPM frame with wires is placed in the vacuum chamber made of titanium, which has good vacuum characteristics and low activation. In order to avoid unnecessary irradiation of the wires, the frame can retract and moves like pendulum motion. During the profile measurement, the beam loss is caused by the interaction at wires, which can be utilized to calibrate the beam loss monitors as well. When the amount of the beam loss exceeds the allowable value, the beam is stopped automatically by the interlock system called Machine Protection System (MPS) [65]. The beam loss monitors placed at all of quadrupole magnets, where the width of the beam becomes relative large. For the beam loss monitor, proportional counter and scintillation counter were selected. All of monitor information is acquired through Experimental Physics and Industrial Control System (EPICS) [66] and stored data base for each shots of beam [67]. From the information of beam orbit and width, the beam parameter such as profile on the target is adjusted by the Strategic Accelerator Design code system (SAD) [68] with several shots of beam.

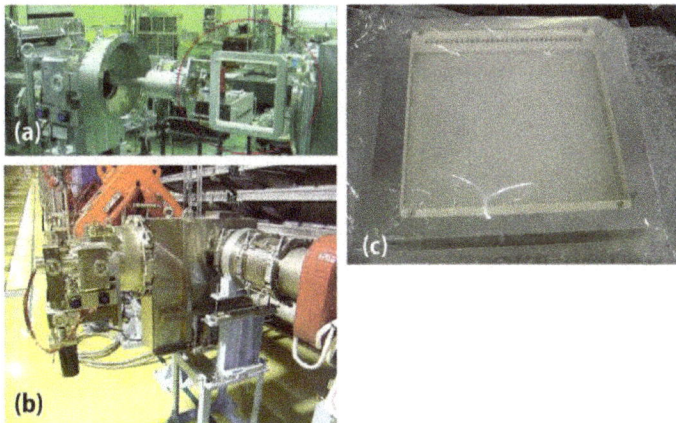

Figure 18. Photographs of Multi Wire Profile Monitor (MWPM) with SiC wires for beam diagnostic along proton beam transport. (**a**) Photograph of MWPC assembled in a chamber; (**b**) Photograph of the chamber containing MWPC installed in the proton beam transport line; (**c**) Photograph of a MWPM frame on which 31 wires of SiC are installed in both horizontal and vertical direction.

5.2.3. Proton Beam Window

To separate ultrahigh vacuum environment in the proton beam transport line from the helium environment in the helium vessel at the target station, a proton beam window (PBW) [69] is placed at 1.8 m upstream of the mercury target. Figure 19 shows a schematic of the PBW. It is made of double walled aluminum alloy of Al5083 with a thickness of 5 mm in total. Al5083 has good characteristics similar to AlMg$_3$ that is used as a safety hull for the 1 MW spallation neutron target at PSI [70], and gives low influence on the scattering of proton beam there. The PBW is replaced every 2 years [69] due to radiation embrittlement. It was estimated that the integral amount of hydrogen produced in Al5083 at the 1 MW proton beam might be equivalent to the total amount produced at the safety hull of PSI. Since the PBW is highly activated after the beam operation, the replacement operation is performed with remote handling tools. Considering the remote handling procedure, an inflatable seal with metal membrane, called pillow seal that was developed by KEK hadron group, is applied to the vacuum seal. Thanks to its good seal performance, very low pressure of less than 10^{-6} Pa has been achieved in the proton beam duct [71]. As shown in Figure 18, radiation shield made of steel with a height of 4 m is connected to the PBW. After the beam operation, the radiation dose rate is lower than 0.2 mSv/h at the flange on the top which is located on the same height as the top of the target station. Access to piping works can be performed by human hands. After removing pipes, the PBW can be plugged out by a crane and pulled into a massive shield cask, and then transferred to the hot cell. In the hot cell, the shielding for the PBW is detached by a remote handing manipulator, and installed on the new PBW. For the reduction of radiation waste volume, water pipes and monitor cables for the PBW are cut by a cutting device placed in the hot cell.

The beam monitors were also assembled to the PBW to observe the characteristics of the beam delivered to the mercury target during the operating period. As shown in Figure 19, the beam monitor, MWPM, was placed at the vacuum side of the PBW. The present SiC wires of MWPC are exposed to the intense proton beams, but it sustained in good conditions until the PBW is replaced. To observe the beam halo at the vicinities of the mercury target, two kinds of beam halo monitors were also placed at the PBW: one is a secondary electron emission type (SEM) to measure relative intensity and the other is a thermocouple type (TC) to measure the intensity quantitatively.

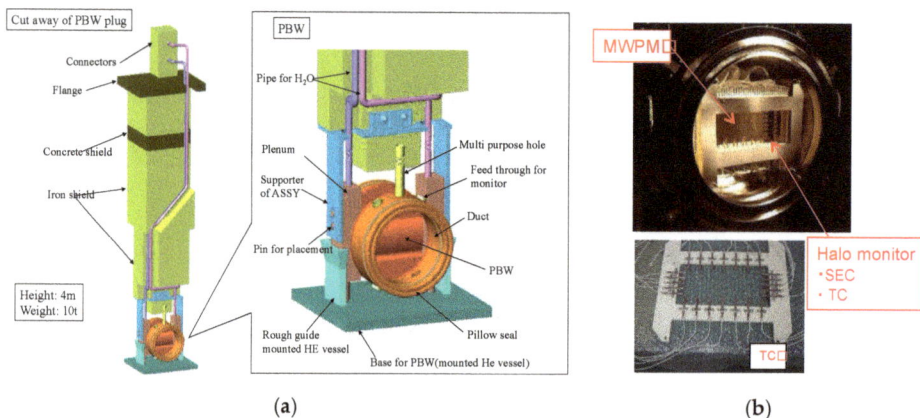

Figure 19. (a) Schematic drawing of proton beam window (PBW); (b) Beam monitor of MWPM and halo monitor consisting of a secondary electron emission counter (SEC) and thermo couples (TC) placed at the PBW.

5.3. Non-Linear Beam Optics for Mitigation of Cavitation Erosion at the Target Vessel

As increasing the beam power, the cavitation erosion at the mercury target vessel due to the proton beam pulse becomes prominent issues for the lifetime of the target vessel. It is reported that the cavitation erosion at the mercury target vessel is correlated with the proton beam power [72] and the peak current density [22,73]. By changing field of the quadrupole magnet placed in front of the target to enlarge the beam size, the density can be easily decreased, while it causes the increase of heat deposition at the vicinities of the target such as shield, reflector and moderator. Since the cavitation erosion is caused in very short time for each beam shot, an ordinary scanning technique such as pulse bending magnet to obtain flat shape on average is useless to mitigate the erosion. To decrease peak current density sustaining low intensity at the vicinities, nonlinear beam optics with octupole magnets has been developed. Two octupole magnets were placed at upstream of the muon production target (MTG). Due to high order of magnetic field, the beam is shaped to flat distribution [74]. Figure 20 compares the beam shape obtained with nonlinear optics with the experimental results. It is shown that the calculated result is in good agreement with the experimental one even if the proton beam is injected on the muon production target. By introducing nonlinear optics, the peak current density can decrease about 30%, which can be thought to mitigate the pitting erosion effectively.

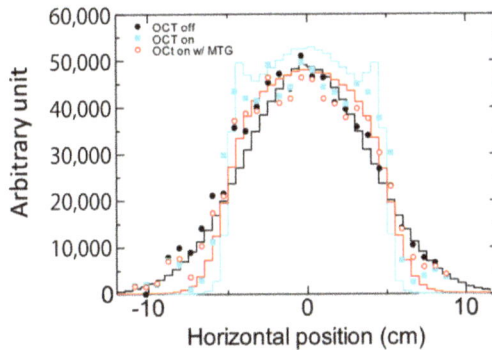

Figure 20. Proton beam profile with nonlinear optics caused by the octupole magnets. Dots stand for the experimental data observed by the MWPM placed at PBW as shown black, cyan and red symbols for results linear optics, nonlinear optics and nonlinear optics with muon production target denoted as muon production target (MTG), respectively. Red, cyan and black lines show calculations with same colors of the experiment.

6. Concluding Remarks

The pulsed spallation neutron source of J-PARC received its very first proton beam from the accelerators in May 2008, and commenced user operation in December 2008. It was measured at a low power operation that a remarkably high neutron intensity of 4.5×10^{12} n/cm^2/s/sr could be emitted from the coupled moderator surface for the 1 MW operation, and a superior resolution of $\Delta d/d = 0.035\%$ was achieved at the beamline BL08 with the poisoned moderator, where d is the lattice-spacing of the Si-400 reflection at 4.628 Å^{-1}. At 300 kW operation, the promising result the pressure wave generated in mercury, could be mitigated to one-third when gas micro-bubbles injection was obtained. In January 2015, the neutron production mercury target received the 1 MW equivalent proton beam pulse for the first time. The beam power for the user program remained still 500 kW so far. Further efforts will be made towards the goal to operate the mercury target system with a beam power of 1 MW.

Acknowledgments: The authors would like to thank for Yujiro Ikeda for his leadership of the neutron source team from the beginning of this work in the late 1990s to until he retired in 2015. Special thanks are due to

Noboru Watanabe for his strong support throughout the execution of this work. His enthusiasm, insights and experience was essential for our design of the high-performance target-moderator-reflector system. The authors also appreciate Guenter Bauer, Tim Broome, Jack Carpenter, Hayo Heyck, Hiroaki Kurishita and Tom McManamy for t plenty of useful advice as our formal technical advisory committee. Their advice helped us to make decisions on the component design. Dave Felde, Bob Sangrey, Bernie Riemer and Mark Wendel are also acknowledged for their useful discussions and collaboration on the pressure wave mitigation in the mercury target. The authors are grateful to Takashi Kato for his leadership in the facility construction period and his support until he retired in 2015 and also to Shin-ichi Sakamoto for his significant contribution to the proton beam transport system from the beginning of design until he retired in 2013. They are indebted to Hideki Tatsumoto for his substantial contribution to the design of cryogenic hydrogen system. Finally, the authors also acknowledge the entire neutron source section team, persons involved in the design and construction of this spallation neutron source for their contributions.

Author Contributions: The draft of Sections 1 and 6 was prepared by H.T.; Section 2 by K.H., H.K., T.N. T.W. and M.F.; Section 3 by M.T., M.O., and M.H.; Section 4 by T.A.; Section 5 by S.-I.M., respectively according to their expertizes and primary fields of contribution to the Pulsed Spallation Neutron Source; Revision of the entire draft for integration into the final production was made by H.T.

Conflicts of Interest: The authors declare no conflicts of interest.

References

1. Sakamoto, S. Technical design report of spallation neutron source facility in J-PARC. In *Neutron Source Section*; JAEA-Technology 2011-035; Japan Atomic Energy Agency: Tokai, Japan, 2012.

2. Sakamoto, S.; Meigo, S.; Fujimori, H.; Harada, M.; Konno, C.; Kasugai, Y.; Kai, T.; Miyake, Y.; Ikeda, Y. Advanced design of high-intensity beam transport line in J-PARC. *Nucl. Instrum. Methods Phys. Res. A* **2006**, *562*, 638–641. [CrossRef]

3. Meigo, S.; Noda, F.; Ishikura, S.; Futakawa, M.; Sakamoto, S.; Ikeda, Y. Evaluation of the 3-GeV proton beam profile at the spallation target of the JSNS. *Nucl. Instrum. Methods Phys. Res. A* **2006**, *562*, 569–572. [CrossRef]

4. Meigo, S.; Ooi, M.; Kai, T.; Ono, T.; Ikezaki, K.; Haraguchi, T.; Fujimori, H.; Sakamoto, S. Beam commissioning for neutron and muon facility at J-PARC. *Nucl. Instrum. Methods Phys. Res. A* **2009**, *600*, 41–43. [CrossRef]

5. Yamazaki, Y. *Accelerator Group JAERI/KEK Joint Project Team, Accelerator Technical Design Report for High-Intensity Proton Accelerator Facility Project, J-PARC*; KEK-Report 2002-13; JAERI-Tech 2003-044; Japan Atomic Energy Agency: Tokai, Ibaraki, Japan, 2003.

6. Ohmori, C.; Ezura, E.; Hashimoto, Y.; Mori, Y.; Schnase, A.; Takagi, A.; Uesugi, T.; Yoshii, M.; Tamura, F.; Yamamoto, M. High field gradient cavity for JAERI-KEK joint project. In Proceedings of the 8th European Particle Accelerator Conference (EPAC 2002), Paris, France, 3–7 June 2002; Laclare, J.-L., Ed.; The European Physical Society Interdivisional Group on Accelerators (EPS-IGA) and CERN: Geneva, Switzerland, 2002; pp. 257–259.

7. Tamura, M.; Maekawa, F. *3-Dimensional Shielding Design for a Spallation Neutron Source Facility in the High Intensity Proton Accelerator Project*; JAERI-Tech 2003-010; Japan Atomic Energy Agency: Tokai, Ibaraki, Japan, 2003. (In Japanese)

8. Teshigawara, M.; Kinoshita, H.; Wakui, T.; Meigo, S.; Seki, M.; Ito, M.; Suzuki, T.; Ikezaki, K.; Maekawa, F.; Futakawa, M.; et al. *Maintenance of Used Components in Spallation Neutron Source, Moderator Reflector and Proton Beam Window*; JAEA-Technology 2012-024; Japan Atomic Energy Agency: Tokai, Ibaraki, Japan, 2012.

9. Haines, J.R.; McManamy, T.J.; Gabriel, T.A.; Battle, R.E.; Chipley, K.K.; Crabtree, J.A.; Jacobs, L.L.; Lousteau, D.C.; Rennich, M.J.; Riemer, B.W. Spallation neutron source target station design, development and commissioning. *Nucl. Instrum. Methods Phys. Res. A* **2014**, *764*, 94–115. [CrossRef]

10. Home page of European Spallation Source. Available online: https://europeanspallationsource.se/european-spallation-source (accessed on 25 July 2017).

11. Kaminaga, M.; Terada, A.; Haga, K.; Kinoshita, H.; Ishikura, S.; Hino, R. *Study of Integrated Structure of Mercury Target Container with Safety Hull*; JAERI-Tech 2000-076; Japan Atomic Energy Agency: Tokai, Ibaraki, Japan, 2001.

12. Futakawa, M.; Kogawa, H.; Hino, R. Measurement of dynamics response of liquid metal subjected to uniaxial strain wave. *J. Phys. IV Fr.* **2000**, *10*, Pr9-237–Pr9-242. [CrossRef]

13. Futakawa, M.; Naoe, T.; Kogawa, H.; Tsai, C.C.; Ikeda, Y. Pitting damage formation up to over 10 million cycles off-line test by MIMTM. *J. Nucl. Sci. Technol.* **2003**, *40*, 895–904. [CrossRef]

14. Naoe, T.; Futakawa, M. Pressure wave-induced cavitation erosion in narrow channel of stagnant mercury. *Trans. JSME* **2014**, *80*, fe0025-1-12. (In Japanese) [CrossRef]

15. Okita, K.; Takagi, S.; Matsumoto, Y. Propagation of pressure waves, caused by a thermal shock, in liquid metals containing gas bubbles. *J. Fluid Sci. Technol.* **2008**, *3*, 116–128. [CrossRef]

16. Futakawa, M.; Kogawa, H.; Hasegawa, S.; Naoe, T.; Ida, M.; Haga, K.; Wakui, T.; Tanaka, N.; Matsumoto, Y.; Ikeda, Y. Mitigation technologies for damage induced by pressure waves in high-power mercury spallation neutron sources (II) bubbling effect to reduce pressure wave. *J. Nucl. Sci. Technol.* **2008**, *45*, 1041–1048. [CrossRef]

17. Kogawa, H.; Naoe, T.; Kyotoh, H.; Haga, K.; Kinoshita, H.; Futakawa, M. Development of microbubble generator for suppression of pressure waves in mercury target of spallation source. *J. Nucl. Sci. Technol.* **2015**, *52*, 1461–1469. [CrossRef]

18. Kinoshita, H.; Haga, K.; Kaminaga, M.; Hino, R. Experiments on mercury circulation system for spallation neutron target. *J. Nucl. Sci. Technol.* **2004**, *41*, 376–384. [CrossRef]

19. Kogawa, H.; Haga, K.; Wakui, T.; Futakawa, M. Development on mercury pump for JSNS. *Nucl. Instrum. Methods Phys. Res. A* **2009**, *600*, 97–99. [CrossRef]

20. Futakawa, M.; Kogawa, H.; Hasegawa, S.; Ikeda, Y.; Riemer, B.; Wendel, M.; Haines, J.; Bauer, G.; Naoe, T.; Okita, K.; et al. Cavitation damage prediction for spallation target vessels by assessment of acoustic vibration. *J. Nucl. Mater.* **2008**, *377*, 182–188. [CrossRef]

21. Teshigawara, M.; Wakui, T.; Naoe, T.; Kogawa, H.; Maekawa, F.; Futakawa, M.; Kikuchi, K. Development of JSNS target vessel diagnosis system using laser Doppler method. *J. Nucl. Mater.* **2010**, *398*, 238–243. [CrossRef]

22. Kogawa, H.; Naoe, T.; Futakawa, M.; Haga, K.; Wakui, T.; Harada, M.; Takada, H. Mitigation technologies for damage induced by pressure waves in high-power mercury spallation neutron source (IV)—Measurement on pressure wave response and microbubble effect on mitigation in JSNS mercury target. *J. Nucl. Sci. Technol.* **2017**, *54*, 733–741. [CrossRef]

23. Kai, T.; Kasugai, Y.; Ooi, M.; Kogawa, H.; Haga, K.; Kinoshita, H.; Seki, M.; Harada, M. Experiences on radioactivity handling for mercury target system in MLF/J-PARC. *Prog. Nucl. Sci. Technol.* **2014**, *4*, 380–383. [CrossRef]

24. Teshigawara, M.; Watanabe, N.; Takada, H.; Nakashima, H.; Nagao, T.; Oyama, Y.; Kosako, K. *Neutronic Studies of Bare Targets for JAERI 5 MW Pulsed Spallation Neutron Source*; JAERI-Research 99-010; Japan Atomic Energy Agency: Tokai, Ibaraki, Japan, 1999.

25. Teshigawara, M.; Harada, M.; Watanabe, N.; Kai, T.; Sakata, A.; Ikeda, Y.; Ooi, M. Proton energy dependence of slow neutron intensity. In Proceedings of the Fifteenth Meeting of the International Collaboration on Advanced Neutron Sources ICANS-XV, Tsukuba, Japan, 6–9 November 2000; Suzuki, J., Itoh, S., Eds.; Japan Atomic Energy Research Institute: Ibaraki, Japan, 2001; pp. 835–847.

26. Broome, T.A.; Hogston, J.R.; Holding, M.; Howells, W.S. The isis methane moderator. In Proceedings of the Twelfth Meeting of International Collaboration on Advanced Neutron Sources, Abingdon, Oxfordshire, UK, 24–28 May 1993; Steigenberger, U., Broome, T., Rees, G., Soper, A., Eds.; Rutherford Appleton Laboratory: Oxfordshire, UK, 1994; pp. T156–T163.

27. Scott, T.L.; Carpenter, J.M.; Miller, M.E. The development of solid methane neutron moderators at the intense pulsed neutron source facility of Argonne National Laboratory. In Proceedings of the International Workshop on Cold Moderators for Pulsed Neutron Sources, Argonne National Laboratory, Lemont, IL, USA, 29 September–2 October 1997; Carpenter, J.M., Iverson, E.B., Eds.; 1997; pp. 299–304. Available online: http://www.osti.gov/scitech/biblio/12385/ (accessed on 25 July 2017).

28. Ishikawa, Y.; Ikeda, S.; Watanabe, N.; Kondo, K.; Inoue, K.; Kiyanagi, Y.; Iwasa, H.; Tsuchihashi, K. Grooved cold moderator at KENS. In Proceedings of the Seventh Meeting of the International Collaboration on Advanced Neutron Sources ICANS-VII, Chalk River, Ontario, ON, Canada, 13–16 September 1983; Schriber, O.S., Ed.; Chalk River Nuclear Laboratories: Ontario, ON, Canada, 1984; pp. 230–235.

29. MacFarlane, R.E. *New Thermal Neutron Scattering Files for ENDF/B-VI Release 2*; LA-12639-MS; Los Alamos National Laboratory Report: Los Alamos, NM, USA, March 1994.

30. Kiyanagi, Y.; Watanabe, N.; Iwasa, H. Premoderator studies for a coupled liquid-hydrogen moderator in pulsed spallation neutron sources. *Nucl. Instrum. Methods Phys. Res. A* **1994**, *343*, 558–562. [CrossRef]

31. Kai, T.; Teshigawara, M.; Watanabe, N.; Harada, M.; Sakara, H.; Ikeda, Y. Optimization of coupled hydrogen moderator for a short pulse spallation neutron source. *J. Nucl. Sci. Technol.* **2002**, *39*, 120–128. [CrossRef]

32. Harada, M.; Teshigawara, M.; Kai, T.; Sakata, H.; Watanabe, N.; Ikeda, Y. Neutronic optimization of premoderator and reflector for decoupled hydrogen moderator in 1 MW spallation neutron source. *J. Nucl. Sci. Technol.* **2002**, *39*, 827–837. [CrossRef]

33. Watanabe, N.; Harada, M.; Kai, T.; Teshigawara, M.; Ikeda, Y. Optimization of coupled and decoupled moderators for a short pulse spallation source. *J. Neutron Res.* **2013**, *11*, 13–23. [CrossRef]

34. Harada, M.; Teshigawara, M.; Watanabe, N.; Kai, T.; Ikeda, Y. Optimization of poisoned and unpoisoned decoupled moderators in JSNS. In Proceedings of the Sixteenth Meeting of the International Collaboration on Advanced Neutron Sources (ICANS XVI), Düsseldorf-Neuss, Germany, 12–15 May 2003; Mank, G., Conrad, H., Eds.; Forschungszentrum Jülich GmbH: Jülich, Germany, 2003; pp. 697–706.

35. Kai, T.; Harada, M.; Teshigawara, M.; Watanabe, N.; Ikeda, Y. Coupled hydrogen moderator optimization with ortho/para hydrogen ratio. *Nucl. Instrum. Methods Phys. Res. A* **2004**, *523*, 398–414. [CrossRef]

36. Kai, T.; Harada, M.; Teshigawara, M.; Watanabe, N.; Kiyanagi, Y.; Ikeda, Y. Neutronic performance of rectangular and cylindrical coupled hydrogen moderators in wide-angle beam extraction of low-energy neutrons. *Nucl. Instrum. Methods Phys. Res. A* **2005**, *550*, 329–342. [CrossRef]

37. Kai, T.; Harada, M.; Teshigawara, M.; Watanabe, N.; Ikeda, Y. Neutronic study on coupled hydrogen moderator for J-PARC spallation neutron source. In Proceedings of the Sixteenth Meeting of the International Collaboration on Advanced Neutron Sources (ICANS XVI), Düsseldorf-Neuss, Germany, 12–15 May 2003; Mank, G., Conrad, H., Eds.; Forschungszentrum Jülich GmbH: Jülich, Germany, 2003; pp. 657–666.

38. Harada, M.; Watanabe, N.; Teshigawara, M.; Kai, T.; Ikeda, Y. Neutronic studies on decoupled hydrogen moderator for a short-pulse spallation source. *Nucl. Instrum. Methods Phys. Res. A* **2005**, *539*, 345–362. [CrossRef]

39. Harada, M.; Watanabe, N.; Teshigawara, M.; Kai, T.; Kato, T.; Ikeda, Y. Neutronics of a poisoned para-hydrogen moderator for a pulsed spallation neutron source. *Nucl. Instrum. Methods Phys. Res. A* **2007**, *574*, 407–419. [CrossRef]

40. Maekawa, F.; Harada, M.; Oikawa, K.; Teshigawara, M.; Kai, T.; Meigo, S.; Ooi, M.; Sakamoto, S.; Takada, H.; Futakawa, M.; et al. First neutron production utilizing J-PARC pulsed spallation neutron source JSNS and neutronic performance demonstrated. *Nucl. Instrum. Methods Phys. Res. A* **2010**, *620*, 159–165. [CrossRef]

41. Batkov, K.; Takibayev, A.; Zanini, L.; Mezei, F. Unperturbed moderator brightness in pulsed neutron sources. *Nucl. Instrum. Methods Phys. Res. A* **2013**, *729*, 500–505. [CrossRef]

42. Mezei, F.; Zanini, L.; Takibayev, A.; Batkov, K.; Klinkby, E.; Pitcher, E.; Schronfeldt, T. Low dimensional neutron moderators for enhanced source brightness. *J. Neutron Res.* **2014**, *17*, 101–105. [CrossRef]

43. Gallmeier, F.X.; Lu, W.; Riemer, B.W.; Zhao, J.K.; Herwig, K.W. Conceptual moderator studies for the spallation neutron source short-pulse second target station. *Rev. Sci. Instrum.* **2016**, *87*, 063304. [CrossRef] [PubMed]

44. Iverson, E.B.; Carpenter, J.M. Kinetics of irradiated liquid hydrogen. In Proceedings of the 6th International Workshop on Advanced Cold Moderators (ACoM-6), Jülich, Germany, 11–13 September 2002; Conrad, H., Ed.; Forschungszentrum Jülich GmbH: Jülich, Germany, 2002; pp. 51–58.

45. Iverson, E.B.; Carpenter, J.M. Kinetics of irradiated liquid hydrogen. In Proceedings of the Sixteenth Meeting of the International Collaboration on Advanced Neutron Sources (ICANS XVI), Düsseldorf-Neuss, Germany, 12–15 May 2003; Mank, G., Conrad, H., Eds.; Forschungszentrum Jülich GmbH: Jülich, Germany, 2003; pp. 707–718.

46. Ooi, M.; Ogawa, H.; Kamiyama, T.; Kiyanagi, Y. Experimental studies of the effect of the ortho/para ratio on the neutronic performance of a liquid hydrogen moderator for a pulsed neutron source. *Nucl. Instrum. Methods Phys. Res. A* **2011**, *659*, 61–68. [CrossRef]

47. Ooi, M.; Ino, T.; Muhrer, G.; Pitcher, E.J.; Russell, G.J.; Ferguson, P.D.; Iverson, E.B.; Freeman, D.; Kiyanagi, Y. Measurements of the change of neutronic performance of a hydrogen moderator at Manuel Lujan Neutron Scattering Center due to conversion from orth- to para-hydrogen state. *Nucl. Instrum. Methods Phys. Res. A* **2006**, *566*, 699–705. [CrossRef]

48. Teshigawara, M.; Harada, M.; Tatsumoto, H.; Aso, T.; Ohtsu, K.; Takada, H.; Futakawa, M.; Ikeda, Y. Experimental verification of equilibrium para-hydrogen levels in hydrogen moderations irradiated by spallation neutrons at J-PARC. *Nucl. Instrum. Methods Phys. Res. B* **2016**, *368*, 66–70. [CrossRef]

49. Teshigawara, M.; Harada, M.; Saito, S.; Oikawa, K.; Maekawa, F.; Futakawa, M.; Kikuchi, K.; Kato, T.; Ikeda, Y.; Naoe, T.; et al. Development of aluminum (Al5083)-clad ternary Ag-In-Cd alloy for JSNS decoupled moderator. *J. Nucl. Mater.* **2006**, *356*, 300–307. [CrossRef]

50. Harada, M.; Saito, S.; Teshigawara, M.; Kawai, M.; Kukuchi, K.; Watanabe, N.; Ikeda, Y. Silver-indium-cadmium decoupler and liner. In Proceedings of the Sixteenth Meeting of the International Collaboration on Advanced Neutron Sources (ICANS XVI), Düsseldorf-Neuss, Germany, 12–15 May 2003; Mank, G., Conrad, H., Eds.; Forschungszentrum Jülich GmbH: Jülich, Germany, 2003; pp. 677–687.

51. Teshigawara, M.; Harada, M.; Saito, S.; Kikkuchi, K.; Kogawa, H.; Kawai, M.; Kurishita, H.; Konashi, K. Cladding technique for development of Ag-In-Cd decoupler. *J. Nucl. Mater.* **2005**, *343*, 154–162. [CrossRef]

52. Kikuchi, K.; Teshigawara, M.; Harada, M.; Saito, S.; Maekawa, F.; Futakawa, M.; Ishigaki, T. A challenge for a high-resolution Ag-In-Cd decoupler in intensified short-pulsed neutron source. *Mater. Sci. Forum* **2010**, *652*, 92–98. [CrossRef]

53. Harada, M.; Teshigawara, M.; Maekawa, F.; Futakawa, M. Study on low activation material for MW-Class spallation neutron sources. *J. Nucl. Mater.* **2010**, *398*, 93–99. [CrossRef]

54. Ooi, M.; Teshigawara, M.; Wakui, T.; Nishi, T.; Harada, M.; Maekawa, F.; Futakawa, M. Development status of low activation ternary Au-In-Cd alloy decoupler for a MW class spallation neutron source: 1st Production of Au-In-Cd alloy. *J. Nucl. Mater.* **2012**, *431*, 218–223. [CrossRef]

55. Ooi, M.; Teshigawara, M.; Kai, T.; Harada, M.; Maekawa, F.; Futakawa, M.; Hashimoto, E.; Segawa, M.; Kureta, M.; Tremsin, A.; et al. Neutron resonance imaging of a Au-In-Cd alloy for the JSNS. *Phys. Procedia* **2013**, *43*, 337–342. [CrossRef]

56. Ooi, M.; Teshigawara, M.; Harada, M.; Naoe, T.; Maekawa, F.; Kasugai, Y. Development of Au-In-Cd decoupler by a hot isostatic pressing (HIP) technique for short pulsed spallation neutron source. *J. Nucl. Mater.* **2014**, *450*, 117–122. [CrossRef]

57. Kamiyama, T.; Oikawa, K. Powder diffractometer at J-PARC. In Proceedings of the Sixteenth Meeting of the International Collaboration on Advanced Neutron Sources (ICANS XVI), Dusseldorf-Neuss, Germany, 12–15 May 2003; Mank, G., Conrad, H., Eds.; Forschungszentrum Jülich GmbH: Jülich, Germany, 2003; pp. 309–314.

58. Aso, T.; Tatsumoto, H.; Hasegawa, S.; Ushijima, I.; Ohtsu, K.; Kato, T.; Ikeda, Y. Design result of the cryogenic hydrogen circulation system for 1 MW pulse spallation neutron source (JSNS) in J-PARC. *AIP Conf. Proc.* **2006**, *823*, 763–770. [CrossRef]

59. Aso, T.; Tatsumoto, H.; Hasegawa, S.; Ohtsu, K.; Uehara, T.; Kawakami, Y.; Sakurayama, H.; Maekawa, F.; Futakawa, M.; Ushijima, I. Commissioning of the cryogenic hydrogen system in J-PARC: Preliminary operation by helium gas. In Proceedings of the Twenty-Second International Cryogenic Engineering Conference and International Cryogenic Materials Conference 2008, Seoul, Korea, 21–25 July 2008; Korea Institute of Applied Superconductivity and Cryogenics: Seoul, Korea, 2009; pp. 741–746.

60. Tatsumoto, H.; Aso, T.; Kato, T.; Ohtsu, K.; Hasegawa, S.; Maekawa, F.; Futakawa, M. Commissioning of the cryogenic hydrogen system in J-PARC: First cool-down operation with helium. In Proceedings of the Twenty-Second International Cryogenic Engineering Conference and International Cryogenic Materials Conference 2008, Seoul, Korea, 21–25 July 2008; Korea Institute of Applied Superconductivity and Cryogenics: Seoul, Korea, 2009; pp. 717–722.

61. Miyake, Y.; Shimomura, K.; Kawamura, N.; Strasser, P.; Makimura, S.; Koda, A.; Fujimori, H.; Nakahara, K.; Kadono, R.; Kato, M.; et al. Birth of an intense pulsed muon source, J-PARC MUSE. *Physica B* **2009**, *404*, 957–961. [CrossRef]

62. Higemoto, W.; Kadono, R.; Kawamura, N.; Koda, A.; Kojima, K. M.; Makimura, S.; Matoba, S.; Miyake, Y.; Shimomura, K.; Strasser, P. Materials and Life Science Experimental Facility at the Japan Proton Accelerator Research Complex IV: The Muon Facility. *Quantum Beam Sci.* **2017**, *1*, 11. [CrossRef]

63. Fujimori, H.; Kawamura, N.; Meigo, S.; Strasser, P.; Nakahara, K.; Miyake, Y. Radiation resistant magnets for the J-PARC muon facility. *Nucl. Instrum. Methods Phys. Res. A* **2009**, *600*, 170–172. [CrossRef]

64. Meigo, S.; Ooi, M.; Ikezaki, K.; Akutsu, A.; Sakamoto, S.; Futakawa, M. Development of profile monitor system for high intense spallation neutron source. In Proceedings of the 1st International Beam Instrumentation Conference, IBIC 2012, Tsukuba, Japan, 1–4 October 2012; Mitsuhashi, T., Shirakawa, A., Eds.; JACow: Geneva, Switzerland, 2013; pp. 227–231. Available online: http://epaper.kek.jp/IBIC2012/papers/mopb68.pdf/ (accessed on 25 July 2017).

65. Sakai, K.; Kai, T.; Ooi, M.; Watanabe, A.; Nakatani, T.; Higemoto, W.; Meigo, S.; Sakamoto, S.; Takada, H.; Futakawa, M. Operation status of interlock system of Materials and Life Science Experimental Facility (MLF) in J-PARC. *Prog. Nucl. Sci. Technol.* **2014**, *4*, 264–267. [CrossRef]

66. Home Page of Experimental Physics and Industrial Control System (EPICS). Available online: http://www.aps.anl.gov/epics/ (accessed on 5 July 2017).

67. Ooi, M.; Kai, T.; Meigo, S.; Kinoshita, H.; Sakai, K.; Sakamoto, S.; Kaminaga, M.; Katoh, T.; Katoh, T. Development of beam monitor DAQ system for 3NBT at J-PARC. In Proceedings of the 10th ICALEPCS International Conference on Accelerator and Large Experiment Physical Control System, Geneva, Switzerland, 10–14 October 2005; PO1.024. pp. 1–6. Available online: http://accelconf.web.cern.ch/accelconf/ica05/proceedings/pdf/P1_024.pdf/ (accessed on 25 July 2017).

68. Home Page of Strategic Accelerator Design (SAD). Available online: http://acc-physics.kek.jp/SAD/ (accessed on 25 July 2017).

69. Meigo, S.; Ooi, M.; Harada, M.; Kinoshita, H.; Akutsu, A. Radiation damage and lifetime estimation of the proton beam window at the Japan Spallation Neutron Source. *J. Nucl. Mater.* **2014**, *450*, 141–146. [CrossRef]

70. Dai, Y.; Hamaguchi, D. Mechanical properties and microstructure of AlMg$_3$ irradiated in SINQ Target-3. *J. Nucl. Mater.* **2005**, *343*, 184–190. [CrossRef]

71. Saito, Y.; Naito, F.; Kubota, C.; Meigo, S.; Fujimori, H.; Ogiwara, N.; Kamiya, J.; Kinsho, M.; Kabeya, Z.; Kubo, T.; et al. Material and surface processing in J-PARC vacuum system. *Vacuum* **2012**, *86*, 817–821. [CrossRef]

72. McClintock, D.A.; Riemer, B.W.; Ferguson, P.D.; Carroll, A.J.; Dayton, M.J. Initial observations of cavitation-induced erosion of liquid metal spallation target vessels at the Spallation Neutron Source. *J. Nucl. Mater.* **2012**, *431*, 147–159. [CrossRef]

73. Naoe, T.; Xiong, Z.; Futakawa, M. Gigacycle fatigue behavior of austenitic stainless steels used for mercury target vessels. *J. Nucl. Mater.* **2016**, *468*, 331–338. [CrossRef]

74. Meigo, S.; Ooi, M.; Ikezaki, K.; Akutsu, A. Beam flattening system based on non-linear optics for high power spallation neutron target at J-PARC. In Proceedings of the 5th International Particle Accelerator Conference, (IPAC2014), Dresden, Germany, 15–20 June 2014; Petit-Jean-Genaz, C., Arduini, G., Michel, P., Schaa, V.R.W., Eds.; JACow: Geneva, Switzerland, 2014; pp. 896–898. Available online: http://accelconf.web.cern.ch/AccelConf/IPAC2014/papers/mopri116.pdf/ (accessed on 25 July 2017).

quantum beam science

MDPI

Review

Materials and Life Science Experimental Facility (MLF) at the Japan Proton Accelerator Research Complex II: Neutron Scattering Instruments

Kenji Nakajima [1,*], Yukinobu Kawakita [1], Shinichi Itoh [1,2,3], Jun Abe [4], Kazuya Aizawa [1], Hiroyuki Aoki [1], Hitoshi Endo [1,2,3], Masaki Fujita [5], Kenichi Funakoshi [4], Wu Gong [6], Masahide Harada [1], Stefanus Harjo [1], Takanori Hattori [1], Masahiro Hino [7], Takashi Honda [1,2], Akinori Hoshikawa [8], Kazutaka Ikeda [1,2,3], Takashi Ino [1,2,3], Toru Ishigaki [8], Yoshihisa Ishikawa [1,2,3], Hiroki Iwase [4], Tetsuya Kai [1], Ryoichi Kajimoto [1], Takashi Kamiyama [1,2,3], Naokatsu Kaneko [1,2], Daichi Kawana [9], Seiko Ohira-Kawamura [1], Takuro Kawasaki [1], Atsushi Kimura [1,10], Ryoji Kiyanagi [1], Kenji Kojima [1,2,3], Katsuhiro Kusaka [8], Sanghyun Lee [1,2], Shinichi Machida [4], Takatsugu Masuda [9], Kenji Mishima [1,2,3], Koji Mitamura [11,12], Mitsutaka Nakamura [1], Shoji Nakamura [1,10], Akiko Nakao [4], Tatsuro Oda [7], Takashi Ohhara [1], Kazuki Ohishi [4], Hidetoshi Ohshita [1,2], Kenichi Oikawa [1], Toshiya Otomo [1,2,3], Asami Sano-Furukawa [1], Kaoru Shibata [1], Takenao Shinohara [1], Kazuhiko Soyama [1], Jun-ichi Suzuki [4], Kentaro Suzuya [1], Atsushi Takahara [11,12], Shin-ichi Takata [1], Masayasu Takeda [13], Yosuke Toh [1,10], Shuki Torii [1,2], Naoya Torikai [2,14], Norifumi L. Yamada [1,2,3], Taro Yamada [8], Dai Yamazaki [1], Tetsuya Yokoo [1,2,3], Masao Yonemura [1,2,3] and Hideki Yoshizawa [9]

[1] Materials & Life Science Division, J-PARC Center, Tokai, Ibaraki 319-1195, Japan;
yukinobu.kawakita@j-parc.jp (Y.K.); shinichi.itoh@kek.jp (S.I.); aizawa.kazuya@jaea.go.jp (K.A.);
hiroyuki.aoki@j-parc.jp (H.A.); hitendo@post.j-parc.jp (H.E.); harada.masahide@jaea.go.jp (M.Ha.);
stefanus.harjo@j-parc.jp (S.H.); hattori.takanori@jaea.go.jp (T.Ha.); takhonda@post.kek.jp (T.Ho.);
kikeda@post.j-parc.jp (K.I.); takashi.ino@kek.jp (T.In.); yoshihisa.ishikawa@j-parc.jp (Y.I.);
tetsuya.kai@j-parc.jp (T.Kai); ryoichi.kajimoto@j-parc.jp (R.Ka.); takashi.kamiyama@kek.jp (T.Kam.);
naokatsu.kaneko@kek.jp (N.K.); seiko.kawamura@j-parc.jp (S.O.-K.); takuro.kawasaki@j-parc.jp (T.Kaw.);
kimura.atsushi04@jaea.go.jp (A.K.); ryoji.kiyanagi@j-parc.jp (R.Ki.); kenji.kojima@kek.jp (K.Ko.);
lee@post.j-parc.jp (S.L.); kenji.mishima@kek.jp (Ke.M.); mitsutaka.nakamura@j-parc.jp (M.N.);
nakamura.shoji@jaea.go.jp (S.N.); takashi.ohhara@j-parc.jp (T.Oh.); ohshita@post.kek.jp (H.O.);
kenichi.oikawa@j-parc.jp (K.Oi.); toshiya.otomo@kek.jp (T.Ot.); sanoasa@post.j-parc.jp (A.S.-F.);
shibata.kaoru@jaea.go.jp (K.Sh.); takenao.shinohara@j-parc.jp (T.S.); soyama.kazuhiko@jaea.go.jp (K.So.);
suzuya.kentaro@jaea.go.jp (K.Su.); takata.shinichi@jaea.go.jp (S.Ta.); toh.yosuke@jaea.go.jp (Y.T.);
torii@post.kek.jp (S.To.); norifumi.yamada@kek.jp (N.L.Y.); yamazaki.dai@jaea.go.jp (D.Y.);
tetsuya.yokoo@kek.jp (T.Yo.); masao.yonemura@kek.jp (M.Y.)

[2] Institute of Materials Structure Science, High Energy Accelerator Research Organization, Tsukuba, Ibaraki 305-0801, Japan; ntorikai@chem.mie-u.ac.jp

[3] Department of Materials Structure Science, SOKENDAI (the Graduate University for Advanced Studies), Tokai, Naka, Ibaraki 319-1106, Japan

[4] Neutron Science and Technology Center, Comprehensive Research Organization for Science and Society, Tokai, Naka, Ibaraki 319-1106, Japan; j_abe@cross.or.jp (J.A.); k_funakoshi@cross.or.jp (K.F.);
h_iwase@cross.or.jp (H.I.); s_machida@cross.or.jp (S.M.); a_nakao@cross.or.jp (A.N.);
k_ohishi@cross.or.jp (K.Oh.); j_suzuki@cross.or.jp (J.S.)

[5] Institute for Materials Research, Tohoku University, Katahira, Aoba-ku, Sendai 980-5877, Japan;
fujita@imr.tohoku.ac.jp

[6] Center for Elements Strategy Initiative for Structural Materials, Kyoto University, Yoshida-honmachi, Sakyo-ku, Kyoto 606-8501, Japan; gong.wu.3x@kyoto-u.ac.jp

[7] Research Reactor Institute, Kyoto University, Kumatori, Sennan, Osaka 590-0494, Japan;
hino@rri.kyoto-u.ac.jp (M.Hi.); t_oda@rri.kyoto-u.ac.jp (T.O.)

8   Frontier Research Center for Applied Atomic Sciences, Ibaraki University, Tokai, Naka, Ibaraki 319-1106,
     Japan; akinori.hoshikawa.eml@vc.ibaraki.ac.jp (A.H.); toru.ishigaki.01@vc.ibaraki.ac.jp (T.Is.);
     katsuhiro.kusaka.1129@vc.ibaraki.ac.jp (K.Ku.); taro.yamada.impromptu@vc.ibaraki.ac.jp (T.Ya.)
9   Neutron Science Laboratory, the Institute for Solid State Physics, the University of Tokyo, Tokai, Naka,
     Ibaraki 319-1106, Japan; kawana@issp.u-tokyo.ac.jp (D.K.); masuda@issp.u-tokyo.ac.jp (T.M.);
     yoshi@issp.u-tokyo.ac.jp (H.Y.)
10   Nuclear Science and Engineering Center, Japan Atomic Energy Agency, Tokai, Naka, Ibaraki 319-1195, Japan
11   JST ERATO, Takahara Soft Interfaces Project, Fukuoka, Fukuoka 819-0395, Japan;
     mitamura@omtri.or.jp (Ko.M.); takahara@cstf.kyushu-u.ac.jp (A.T.)
12   Institute for Materials Chemistry and Engineering, Kyushu University, Fukuoka, Fukuoka 819-0395, Japan
13   Materials Sciences Research Center, Japan Atomic Energy Agency, Tokai, Naka, Ibaraki 319-1195, Japan;
     takeda.masayasu@jaea.go.jp
14   Graduate School of Regional Innovation Studies, Mie University, Tsu, Mie 514-8507, Japan
*   Correspondence: kenji.nakajima@j-parc.jp; Tel.: +81-29-284-3195

Received: 28 June 2017; Accepted: 7 November 2017; Published: 17 November 2017

Abstract: The neutron instruments suite, installed at the spallation neutron source of the Materials and Life Science Experimental Facility (MLF) at the Japan Proton Accelerator Research Complex (J-PARC), is reviewed. MLF has 23 neutron beam ports and 21 instruments are in operation for user programs or are under commissioning. A unique and challenging instrumental suite in MLF has been realized via combination of a high-performance neutron source, optimized for neutron scattering, and unique instruments using cutting-edge technologies. All instruments are/will serve in world-leading investigations in a broad range of fields, from fundamental physics to industrial applications. In this review, overviews, characteristic features, and typical applications of the individual instruments are mentioned.

Keywords: J-PARC; neutron instruments; inelastic neutron scattering; quasielastic neutron scattering; neutron diffraction; neutron reflectometry; small angle neutron scattering; total neutron scattering; prompt gamma-ray analysis; neutron cross section measurement; neutron imaging

PACS: 28.20.Cz; 28.20.Np; 28.20.Pr; 29.30.Hs; 07.35.+k; 07.05.Hd; 07.05.Kf; 60.; 70.; 80.; 90

1. Introduction

The spallation neutron source at the Materials and Life Science Facility (MLF) is one of the most advanced and one of the most powerful neutron sources in the world. In two experimental halls at MLF (Figure 1), 20 neutron instruments are in operation and one is under commissioning (Table 1). These are various types of instruments to serve various fields of investigations in which neutron scattering techniques can play a significant role.

Each of our instruments is unique, state-of-the-art, and distinguished from other instruments that have existed thus far. They are realized by a combination of the high performance of our neutron source and new technologies.

The characteristics of our neutron source are one of the most important key factors contributing to their uniqueness and the high performance of the instruments. When the Japan Proton Accelerator Research Complex (J-PARC) project was launched, individuals from the Japanese neutron science community got together and intensively discussed possible neutron instruments at the high-intensity spallation neutron source [1]. The results of these discussions became the foundation of the Grand Design, a list of instruments to be built at MLF, which were established by the Japanese Society for Neutron Science. Over the course of the discussions, through dialogue with the neutron source group, the unique features of our neutron source were determined, e.g., (1) all of moderators are cryogenic (supercritical H_2) without ambient ones; (2) the majority of the beam ports needed to view a coupled moderator; and (3) the pulse repetition rate needed to be 25 Hz. In particular, a coupled moderator is one of key features MLF. It was obvious that a fairly high neutron flux was expected to be obtained using the new 1 MW-class spallation source, which we have not yet experienced. However, we were not satisfied with this expectation, and attempted to maximize the beam flux further. The answer to this requirement is that our coupled moderator, which provides a maximum intensity at the neutron source at MLF. The disadvantage of this type of moderator is that the pulse shape becomes rather broad and strongly asymmetric, especially in the lower energy region. In spite of this fact, almost half of the neutron instruments at MLF have a coupled moderator. It should be noted that three of them are chopper spectrometers and one is a near-backscattering spectrometer, for which a sharp pulse source was considered to be indispensable. On the other hand, another aspect of the needed neutron source, which prefers a finer and symmetric pulse shape, resulted in decoupled and decoupled poison moderators. With its fine pulse shape, the new Ag-In-Cd decoupler contributes to the high performance of our high resolution diffractometers. SuperHRPD surpassed the world's finest achieved resolution record, even at the very early stages of MLF. The characteristics of our neutron source allow the neutron instruments to attempt new methods, new technologies, and new devices, as described in the following sections.

In addition to utilizing the advantages of our advanced neutron source, we have been attempting to involve new techniques, as well as new ideas, to maximize the performance of our instruments, some which were considered to be challenging at the time but are now common. From the beginning of the J-PARC project, we decided to employ an event data-recording method, instead of histogram recording, which was the standard method over the long history of pulsed neutron source instruments [2,3]. All software suites, data acquisition systems, and instruments were designed to fully adapt to the event recording method, and this idea enabled us to perform multi-incident energy measurements using chopper spectrometers [4], stroboscopic measurements using diffractometers [5], and even magnetic field imaging using our imaging instruments [6,7]. The pulse manipulation techniques gave us another success; multi-incident energy measurements using a repetition rate multiplication (RRM) technique [8] were realized using 4D-Space Access Neutron Spectrometer (4SEASONS) as the world's first demonstration [4], and this method became quite common for chopper spectrometers at modern pulsed neutron facilities around the world. Pulse-shaping technology using a pulse shaping chopper located near a source was employed at the Cold-Neutron Disk-Chopper Spectrometer AMATERAS [9] around the same time that it was performed at CNCS at the Spallation Neutron Source (SNS) (Oak Ridge National Laboratory, Oak Ridge, TN, USA) [10] and LET at ISIS Neutron and Muon Source (Rutherford Appleton Laboratory, Didcot, UK) [11]. These three chopper spectrometers were the world's first cases of using a pulse-shaping chopper to control resolution. DNA was the world's first example of a practical realization of the wave-length frame multiplication technique, which permits pulse-shaping, even on an indirect geometry spectrometer [12]. Another interesting challenge is the neutron Brillouin scattering (NBS) capability, which is being developed as an option at the High Resolution Chopper Spectrometer (HRC) to overcome the kinematic constraints of neutron spectroscopy [13], and it has been confirmed that the performance is better than those of existing NBS-dedicated instruments. We also attempted to use many other technologies, such as wavelength shifting fiber scintillators, elliptic guide geometry, pulse neutron adapted magnetic lens, etc., which provided a breakthrough for our instruments to realize the performance that would be expected by users. Although some of these breakthroughs are commonplace components in many instruments at many facilities now, they were challenging when we started pioneering work on them.

On the basis of these technological developments, neutron applications have been widely extended by accompanying a kind of phase change. For example, diffraction measurements are trying not only passive measurement with static conditions at specific temperatures, but also active measurements, such as stroboscopic measurements under elevating temperatures passing over phase transitions, alternating external fields, in situ stress measurements under a loading process, and so on. Inelastic scattering measurements have been drastically changed from a pin-point measurement in specific regions of momentum and energy transfers to observations covering an entire range using several energy resolutions. Materials to be measured have gradually changed from pure materials, from which fundamental aspects can be extracted, to composite materials, which can be used in industrial application. Highly-intense beams open new research fields, such as neutron bio-crystallography and high-pressure geo-sciences. The development of software is always key in supporting new experimental techniques and in developing new analysis methods.

Nine years have passed since we produced the first neutron beam at MLF, in May, 2008. Now, 21 of 23 beam ports are presently available. We are always attempting to realize cutting-edge instrumentation. In April 2015, our newest instrument, the Energy Resolved Neutron Imaging System RADEN at BL22, was released in the user program. RADEN is the world's first dedicated energy-resolved imaging instrument, and provides full-scale possibilities for resonance absorption, Bragg edge imaging, and magnetic field imaging. The Village of Neutron Resonance Spin Echo Spectrometers VIN ROSE at BL06 is an ambitious spin echo spectrometer suite composed of a neutron-resonance-type spectrometer and a modulated intensity by zero effort type spin-echo spectrometer. The spectrometer has been partly open to the user program since the latter half of 2017. Moreover, a new chopper spectrometer, POLANO (Polarized Neutron Spectrometer), is under commission at BL23. POLANO is a dedicated polarized neutron inelastic instrument, in which we are employing modern polarization techniques.

Our neutron instruments at MLF are operated by four different groups:

(i)   Spectroscopy group (inelastic instruments)

  4SEASONS: 4D-Space Access Neutron Spectrometer [14]

  HRC: High Resolution Chopper Spectrometer [15,16]

  AMATERAS: Cold-Neutron Disk-Chopper Spectrometer [9]

  POLANO: Polarized Neutron Spectrometer [17–19]

  DNA: Biomolecular Dynamics Spectrometer [12]

  VIN ROSE: Village of Neutron Resonance Spin Echo Spectrometers [20].

(ii)   Crystal-structure group (diffractometers)

  SuperHRPD: Super High-Resolution Powder Diffractometer [21]

  SPICA: Special Environment Powder Diffractometer [22]

  iMATERIA: IBARAKI Materials Design Diffractometer (Versatile Neutron Diffractometer) [23]

  PLANET: High-Pressure Neutron Diffractometer [24]

  TAKUMI: E Engineering Materials Diffractometer [25,26]

  iBIX: IBARAKI Biological Crystal Diffractometer [27,28]

  SENJU: Extreme Environment Single-crystal Neutron Diffractometer [29].

(iii)   Nano-structure group (small-angle neutron scattering instrument, reflectometers, and total scattering instruments)

  TAIKAN: Small- and Wide-angle Neutron Scattering Instrument [30]

  SOFIA: Soft Interface Analyzer [31,32]

  SHARAKU: Polarized Neutron Reflectometer [33]

  NOVA: High-Intensity Total Diffractometer.

(iv) Pulsed neutron application group (other than neutron scattering instruments (beamlines for neutronics studies, fundamental physics, prompt gamma-ray analysis, versatile test port, and neutron imaging))

ANNRI: Accurate Neutron-Nucleus Reaction measurement Instrument [34]

NOP: Neutron Optics and Physic [35]

NOBORU: NeutrOn Beam-line for Observation and Research Use [36]

RADEN: Energy Resolved Neutron Imaging System [37].

In this review, we overview these 21 instruments at MLF, according to these 4 groups. In addition, we review the typical scientific outputs from these instruments.

Figure 1. Photo of experimental halls No. 1 (**left**) and No. 2 (**right**) of the Materials and Life Science Experimental Facility (MLF).

Table 1. Neutron instruments at MLF (as of May 2017).

Beam Line	Moderator	Short Name	Formal Name	Status	Reference
BL01		4SEASONS	4D-Space Access Neutron Spectrometer	Operation (2008~)	[14]
BL02		DNA	Biomolecular Dynamics Spectrometer	Operation (2011~)	[12]
BL03		iBIX	IBARAKI Biological Crystal Diffractometer	Operation (2008~)	[27,28]
BL04	Coupled H_2	ANNRI	Accurate Neutron-Nucleus Reaction Measurement Instrument	Operation (2008~)	[34]
BL05		NOP	Neutron Optics and Fundamental Physics	Operation (2008~)	[35]
BL06		VIN ROSE	Village of Neutron Resonance Spin Echo Spectrometers	Operation (2017~)	[20]

Table 1. *Cont.*

Beam Line	Moderator	Short Name	Formal Name	Status	Reference
BL07			*Vacant*		
BL08	Decoupled Poisoned H_2 (thin)	SuperHRPD	Super High-Resolution Powder Diffractometer	Operation (2008~)	[21]
BL09		SPICA	Special Environment Powder Diffractometer	Operation (2011~)	[22]
BL10		NOBORU	NeutrOn Beam-line for Observation and Research Use	Operation (2008~)	[36]
BL11	Decoupled H_2	PLANET	High-Pressure Neutron Diffractometer	Operation (2013~)	[24]
BL12		HRC	High Resolution Chopper Spectrometer	Operation (2008~)	[15,16]
BL13			*Vacant*		
BL14		AMATERAS	Cold-Neutron Disk-Chopper Spectrometer	Operation (2009~)	[9]
BL15	Coupled H_2	TAIKAN	Small- and Wide-angle Neutron Scattering Instrument	Operation (2011~)	[30]
BL16		SOFIA	Soft Interface Analyzer	Operation (2008~)	[31,32]
BL17		SHARAKU	Polarized Neutron Reflectometer	Operation (2011~)	[33]
BL18		SENJU	Extreme Environment Single-crystal Neutron Diffractometer	Operation (2011~)	[29]
BL19	Decoupled Poisoned H_2 (thick)	TAKUMI	Engineering Materials Diffractometer	Operation (2008~)	[25,26]
BL20		iMATERIA	IBARAKI Materials Design Diffractometer (Versatile Neutron Diffractometer)	Operation (2008~)	[23]
BL21		NOVA	High-Intensity Total Diffractometer	Operation (2008~)	
BL22	Decoupled H_2	RADEN	Energy Resolved Neutron Imaging System	Operation (2015~)	[37]
BL23		POLANO	Polarized Neutron Spectrometer	Commissioning	[17–19]

2. Spectroscopy Group Instruments

The MLF Spectroscopy Group consists of six instruments for inelastic and quasi-elastic neutron scattering in MLF [38]. 4SEASONS, HRC, AMATERAS, and POLANO are direct geometry spectrometers; 4SEASONS, HRC, and POLANO use Fermi choppers to monochromate incident neutron beams, while AMATERAS used a combination of disk choppers. Among these spectrometers, POLANO can be used with polarized neutrons. Its construction was completed recently, and it has been in the commissioning phase since 2017. DNA is a near-back scattering spectrometer with a pulse-shaping chopper. Measurements with micro-eV resolution can be performed using this instrument. Neutron spin echo (NSE) spectrometer VIN ROSE can access even slower dynamics. This spectrometer includes two instruments of the Modulated Intensity with Zero Effort (MIEZE) type and Neutron Resonance Spin Echo (NRSE) type. They are under commissioning and are expected to be open to the user program in 2017. Combinations of these instruments in MLF allows for the study of dynamics in diverse fields, including solid state physics, amorphous materials and liquids, soft and biological matters, as well as industrial applications such as tire rubbers and battery materials (Figure 2).

As for the operational sides of the instruments, AMATERAS is a Japan Atomic Energy Agency (JAEA) beamline operated by JAEA staff, while 4SEASONS and DNA are the Public Beamlines operated by JAEA and the registered institution for public use, Comprehensive Research Organization for Science and Society (CROSS). HRC, POLANO, and VIN ROSE are operated by the High Energy Accelerator Research Organization (KEK) in collaboration with the University of Tokyo, Tohoku University, and Kyoto University, respectively, under KEK Inter-University Research Program. Despite the differences in their operating policies, there is no restriction on user access, and the knowledge obtained by these instruments is shared for improving instrument performance through moderate collaboration among the institutions that run them.

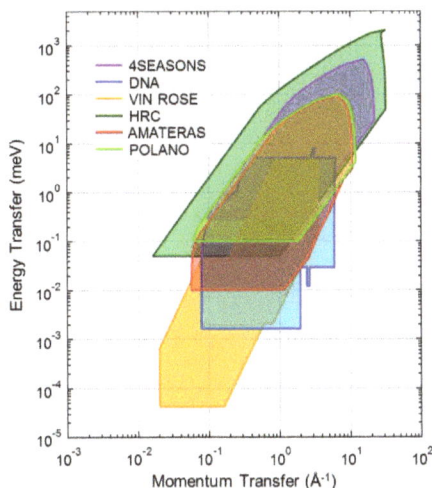

Figure 2. Momentum-energy space covered by six spectrometers in MLF [38].

2.1. 4D-Space Access Neutron Spectrometer, 4SEASONS

4SEASONS is a thermal neutron Fermi chopper spectrometer designed for measurements of dynamics in the 10^0–10^2 meV energy range [14,38]. A schematic view of 4SEASONS is presented in Figure 3a. Characteristic parameters of the spectrometer are listed in Table 2. It is installed at BL01 beam port for viewing the coupled moderator. Neutrons are transported to a sample position located 18 m downstream of the moderator through an elliptically converging straight neutron guide tube coated with supermirrors [39]. The incident neutrons are monochromated by a fast-rotating Fermi chopper positioned 1.7 m upstream of the sample position. In addition, the instrument has a T0 chopper for suppressing fast neutrons and two disk choppers for band definition. Neutrons scattered by a sample are detected over time by the one-dimensional (1D) ^{3}He position-sensitive detectors (PSDs) placed at 2.5 m from the sample position. The angular coverage of the detectors relative to the direct beam ranges from −35° to +91° horizontally and from −25° to +27° vertically. The sample environment devices and detectors are enclosed in a large vacuum chamber without any window separating them, which minimizes background scattering and detector gaps (Figure 3b). To further reduce the background scattering from the sample environment, an oscillating radial collimator is available [40].

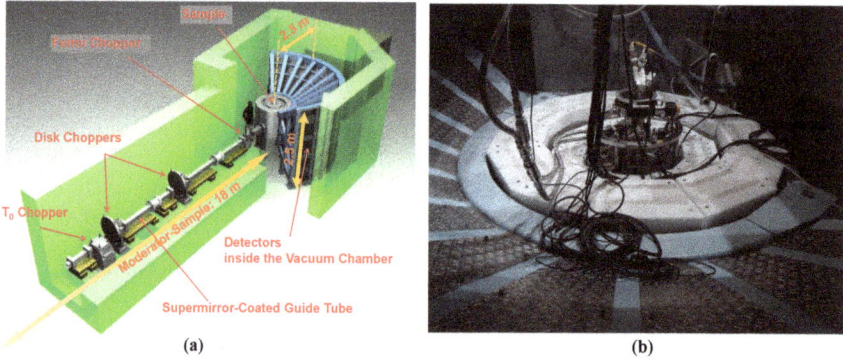

(a)   (b)

Figure 3. (a) Schematic view of 4SEASONS [38]; (b) A cryostat for cooling the sample is attached to the vacuum chamber of 4SEASONS.

Table 2. Specifications of 4SEASONS.

Beamline	BL01
Moderator	Coupled hydrogen moderator
Flight path length	$L_{\text{moderator-sample}} = 18$ m $L_{\text{sample-detector}} = 2.5$ m
Incident energy	5–300 meV
Energy resolution	$\Delta E/E_i \geq 5\%$ FWHM at the elastic line
Detectors	^{3}He 1D- position-sensitive detectors (PSD) (16.4 atm partial pressure) 19 mm diameter, 2500 mm long, 352 (266*) tubes Angular coverage: $-35° - +130°$ ($-35° - +91°$ *) horizontal $-25° - +27°$ vertical (* current values)
Fermi chopper	$L_{\text{moderator-chopper}} = 16.3$ m Long slit package: 2 mm × 100 mm slots, revolution rate ≤ 350 Hz Short slit package: 0.4 mm × 20 mm slots, revolution rate ≤ 600 Hz Radius = 350 mm
Slow disk choppers	Revolution rate = 12.5 Hz or 25 Hz No. 1: $L_{\text{moderator-chopper}} = 9$ m, opening angle = 77° No. 2: $L_{\text{moderator-chopper}} = 12$ m, opening angle = 103°
T0 chopper	$L_{\text{moderator-chopper}} = 8.5$ m Revolution rate = 25 Hz
Beam transport	$m = 3.2$–4 supermirror
Beam size at the sample position	maximum 45 mm × 45 mm, optimum 20 mm × 20 mm, adjustable by motorized slits

Compared to the other chopper spectrometers in MLF, 4SEASONS is designed to supply high thermal neutron flux by relaxing the resolution to observe weak inelastic signals [38]. Typical energy and momentum transfer resolutions under the elastic scattering condition are 6% relative to the incident energy and 1–2% relative to the incident wave number, respectively [14,41]. Another important feature of 4SEASONS is that it can perform measurements at multiple incident energies simultaneously (multi-E_i measurements) [4]. This was achieved by taking advantage of the fact that the Fermi chopper rotates considerably faster than the repetition rate of the neutron source, which is one of the simplest realizations of RRM [42,43]. The practical number of available incident energies is 3–4, which covers one or two orders of magnitude on the energy scale depending on experimental conditions.

These features of the instrument were originally designed for studies of high-critical-temperature (high-T_c) oxide superconductors. Indeed, copper and iron-based superconductors are still major research targets at 4SEASONS [44–49]. Including these materials, strongly correlated electron systems and magnetism dominate 80% of the experimental proposals. Figure 4a shows one of the examples of 4SEASONS data, indicating the excitation spectrum of a single crystal of $Pr_{1.40}La_{0.60}CuO_4$ [45]. Well-defined spin wave excitations emerging from $h = 0.5$ and 1.5 can be seen clearly up to ~300 meV. Another recent trend of experiments at 4SEASONS is four-dimensional (4D) mapping of the energy–momentum space in a single crystal by rotating the crystal. Figure 4b shows an example of the rotating-crystal measurement performed at 4SEASONS. It shows two-dimensional (2D) maps of the 4D phonon spectrum of copper as cuts on the (*H*, *K*, *K*), *H-E*, and *K-E* planes. Thanks to the high neutron flux, such a 4D spectrum can be obtained within a reasonable measurement time (1–3 days). This type of measurement is employed in more than 30% of experiments performed at 4SEASONS.

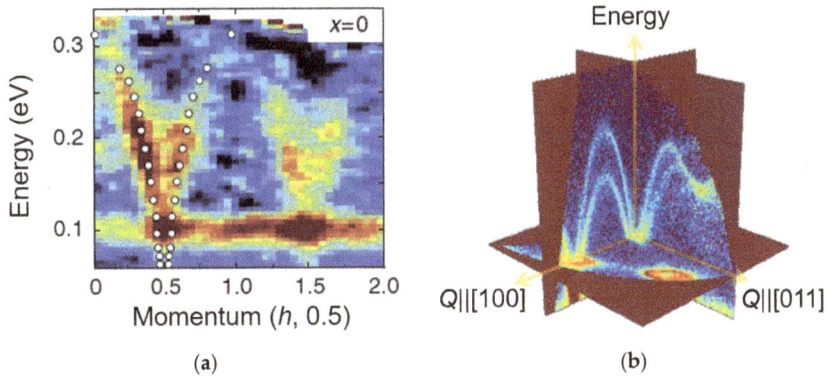

Figure 4. Examples of excitation spectra observed at 4SEASONS. The data are shown as functions of momentum and energy of excitations. Colors indicate neutron-scattering intensities: (a) Magnetic excitation spectrum of $Pr_{1.40}La_{0.60}CuO_4$ [45]; (b) Phonon dispersions of Cu obtained by crystal-rotating measurement.

2.2. High Resolution Chopper Spectrometer, HRC

The High Resolution Chopper Spectrometer (HRC) offers high resolutions and delivers relatively high-energy neutrons for a wide range of studies on the dynamics of materials [13,15,16]: the range of incident neutron energies E_i = 5–2000 meV is available, and especially, by using $E_i \leq$ 300 meV, energy resolution of $\Delta E/E_i$ = 2% can be achieved in the best case. For conventional experiments, the energy resolution is set to $\Delta E/E_i$ = 3–10% to increase the neutron flux. A schematic layout of HRC is illustrated in Figure 5, and the characteristic parameters are listed in Table 3. On the primary flight path of 15 m, a supermirror guide, T0 chopper running up to 100 Hz at 9 m from the neutron source, Fermi chopper at 14 m from the source, and an incident beam collimator system just at the upper stream of the sample are installed. HRC has a detector array of ^{3}He PSDs measuring 2.8 m in length and 19 mm in diameter at 4 m from the sample position, and it covers scattering angles from 3° to 62° for conventional experiments. Moreover, another array of ^{3}He PSDs measuring 0.8 m in length and 12.7 mm in diameter is installed at 5.2 m from the sample position, and it covers scattering angles down to 0.6°. The incident beam collimator system having two collimators composed of slits of vertical sheets of Cd is installed for reducing background noise and one of the two collimators is selected: coarse collimation of 1.5° for conventional experiments using detectors down to 3° and fine collimation of 0.3° for low-angle experiments using detectors down to 0.6°.

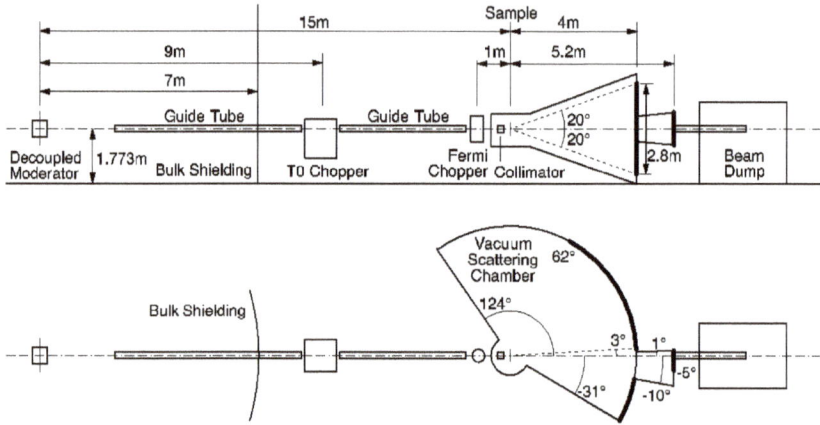

Figure 5. Layout of HRC. Thick lines indicate arrays of ^{3}He PSDs.

Table 3. Specifications of HRC (see [13,15,16] for details).

Beamline	BL12
Moderator	Decoupled hydrogen moderator
Flight path length	$L_{\text{moderator-sample}}$ = 15 m $L_{\text{sample-detector}}$ = 4 m, 5.2 m
Incident energy	5–2000 meV
Energy resolution	$\Delta E/E_i$ = 3–10% (conventional), $\Delta E/E_i \geq 2\%$ (neutron Brillouin scattering (NBS))
Q resolution (designed)	$\Delta Q/k_i \geq 1\%$
Detector coverage (scattering angle)	Horizontal: $-31° - +124°$ (designed) $-31° - +62°$ (current) Vertical: $-20° - +20°$
Detector system	^{3}He 1D-PSD [19.05 mm (diameter), 28,000 mm (length)] (256 tubes are currently installed) for 4 m position ^{3}He 1D-PSD [12.7 mm (diameter), 8000 mm (length)] (68 tubes are currently installed) for 5.2 m position
T0 chopper	Revolution rate : 25, 50, 100 Hz $L_{\text{moderator-chopper}}$ = 9 m
Fermi chopper	Revolution rate: 100–600 Hz $L_{\text{moderator-chopper}}$ = 14 m
Beam transport	m = 3, 3.65 and 4 supermirror
Beam size at sample position	Maximum dimension: 50 mm width × 50 mm height

At HRC, three types of inelastic neutron scattering experiments can be performed: high-resolution experiments in the conventional energy-momentum space, eV neutron spectroscopy, and neutron Brillouin scattering (NBS). In high-resolution experiments in the conventional space, the dynamical structure factor in a spin system can be determined in the full energy-momentum space, and many studies on condensed matter physics can be performed, for instance [50]. The dynamical structure factors of the multiferroic system $NdFe_3(BO_3)_4$, layered nickelate $R_{2-x}Sr_xNiO_4$ (R = La and Nd), and carrier-doped Haldane system $Nd_{2-x}Ca_xBaNiO_5$ were determined over the entire Brillouin zone. As a result, microscopic interactions in these systems could be discussed. In eV neutron spectroscopy, intermultiplet transitions in a skutterudite compound $SmFe_4P_{12}$ with natural Sm were observed successfully by using E_i = 0.5 eV, where the absorption cross-section of natural Sm is the minimum.

We have started observing high-energy magnetic excitations in metallic antiferromagnets such as Cr. When the full power beam is available, we expect to observe electronic excitations [51].

Owing to the kinematic constraints of neutron spectroscopy, to observe NBS, that is, inelastic neutron scattering close to the forward direction, an incident neutron energy in the sub-eV region with high resolution is necessary, and the scattered neutrons need to be detected at very low scattering angles. At HRC, NBS experiments became feasible by reducing the background noise at low angles down to $0.6°$. The principle of NBS is not new [52–54], and the energy–momentum space accessible by using NBS has been extended utilizing higher-energy neutrons in spectrometers such as BRISP at the Institut Laue-Langevin (ILL, Grenoble, France) [55].

To show the feasibility of NBS at HRC, observation of spin waves was demonstrated using a polycrystalline sample of a well-known cubic perovskite ferromagnet, $La_{0.8}Sr_{0.2}MnO_3$: the observed spin wave dispersion relation agreed well with the previous single-crystal result [13]. For a similar cubic perovskite ferromagnet $SrRuO_3$, we found a spin wave gap, as shown in Figure 6a [13]. $SrRuO_3$ is a metallic ferromagnet with transition temperature $T_C = 165$ K. Normally, cubic ferromagnets such as Fe, Ni, and $La_{0.8}Sr_{0.2}MnO_3$ show very weak magnetic anisotropy, and therefore, the spin wave gap is negligibly small. However, $SrRuO_3$ shows a finite spin wave gap. Moreover, $SrRuO_3$ shows that anomalous Hall resistivity is not proportional to magnetization [56]. In $SrRuO_3$, the band structure exhibits Weyl fermions (band crossings) owing to spin orbit interaction, which produces the Berry phase. The Berry curvature of the band crossing takes the functional form of the magnetic field of a monopole in the momentum space, and the fictitious magnetic field of the monopole is the origin of the anomalous Hall effect: the anomalous Hall conductivity σ_{xy} in $SrRuO_3$ is well described by the Berry curvature of the Weyl fermion [56]. To detect the fictitious magnetic field in $SrRuO_3$ with inelastic neutron scattering, we performed NBS experiments [57] because a large single crystal necessary for inelastic neutron scattering had not been synthesized until very recently. Well-defined spin wave peaks were observed, and the dispersion curve was well fitted to $E(Q) = DQ^2 + E_g$ at temperatures (T) below T_C. The spin wave gap E_g showed non-monotonous T dependence and was well explained by the theoretical function $E_g(T) = aM(T)/[1 + bM(T)\sigma_{xy}(T)]$ with adjustable parameters a and b, where $M(T)$ is spontaneous magnetization, as shown in Figure 6b. As mentioned above, σ_{xy} is described by the Berry curvature, and therefore, we showed for the first time that inelastic neutron scattering detects the Berry phase or the fictitious magnetic field of monopoles through σ_{xy}. Until this study, Weyl fermions were discussed in transport phenomena only, in terms of spintronics. We showed that the Berry curvature is an observable of inelastic neutron scattering and that the spin dynamics directly reflects Weyl fermions. This result has revealed a novel connection between the transport and the dynamical magnetic properties through the enhanced spin-orbit coupling effect.

At HRC, many high-resolution experimental studies in the conventional energy-momentum space and eV neutron spectroscopy studies have been performed or are in progress. We realized NBS experiments by using high-energy neutrons with high resolution by successful reduction of background noise at low angles. The NBS option makes HRC different from other inelastic neutron scattering instruments and opens up opportunities for contributing to current science by measuring collective excitations of non-single-crystal samples.

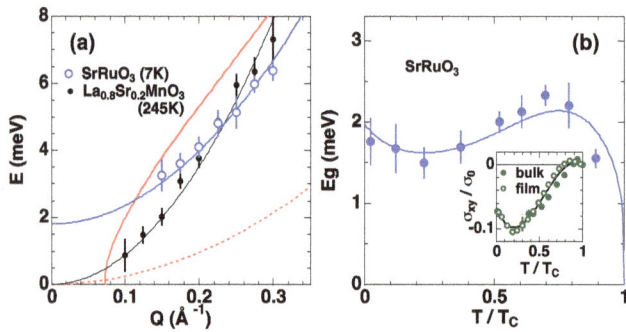

Figure 6. (**a**) Spin wave dispersion relationships of SrRuO$_3$ ($T = 7$ K) and La$_{0.8}$Sr$_{0.2}$MnO$_3$ ($T = 245$ K) observed with NBS at HRC [13]. The black line is a single-crystal result, and the blue line is a fitted curve with $E(Q) = DQ^2 + E_g$. The accessible energy-momentum space in this experiment is below the red line, which is the scan locus of a detector at the scattering angle $\phi = 0.6°$ with $E_i = 100$ meV, and the red dotted line is the upper limit of the accessible space for $\phi \geq 5°$; (**b**) $E_g(T)$ of SrRuO$_3$ [57]. The blue line is a theoretical curve (see text). The inset shows $\sigma_{xy}(T)$ for bulk and film samples, where σ_0 is a constant, and the solid line is an empirical function describing $\sigma_{xy}(T)$ for the analysis of $E_g(T)$.

2.3. Cold-Neutron Disk-Chopper Spectrometer, AMATERAS

AMATERAS (Figure 7) is a multi disk-chopper spectrometer at BL14 [9,38,58]. The spectrometer is designed to carry out inelastic and quasielastic neutron scattering experiments on single crystal, powder, liquid and amorphous samples at neutron energies ranging from cold to sub-thermal. The most characteristic feature of AMATERAS is that the spectrometer employs the pulse-shaping technique, which has also been adapted to other modern cold-neutron chopper spectrometers at pulsed sources, namely, CNCS at SNS [10] and LET at ISIS [11]. One set of fast disk choppers located at the upstream position cuts out the ideal (sharp and symmetric) peak from the source pulse from a coupled moderator at MLF, which has a large intensity but broad and asymmetric pulse shape. By using a pulse-shaping technique and owing to the high peak intensity of the coupled moderator at MLF, AMATERAS realizes high-intensity, fine, and flexible energy resolution measurements [9]. AMATERAS has capability to carry out multi-E_i measurements by using RRM technique that was demonstrated for the first time at MLF [4].

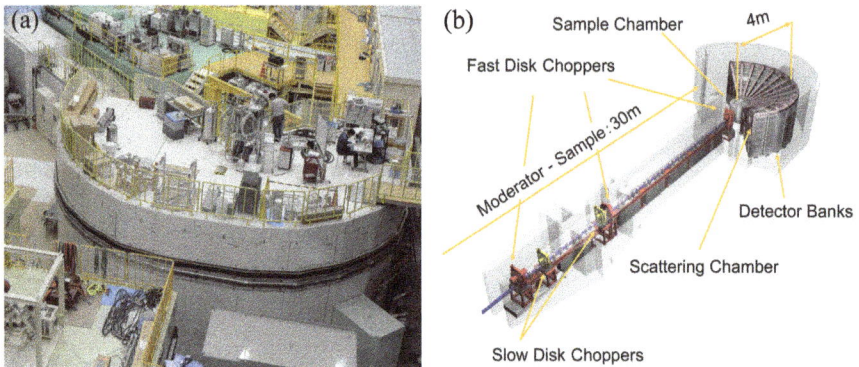

Figure 7. (**a**) Photo and (**b**) schematic view [38] of AMATERAS.

The characteristic parameters of AMATERAS are listed in Table 4. The spectrometer uses the incident energy of 1–80 meV, while the best performance can be achieved at $E_i \leq 20$ meV. The finest energy resolution is $\Delta E / E_i = 1\%$ at $E_i = 20$ meV, and it becomes finer as the incident energy decreases. AMATERAS is equipped with newly developed fast disk choppers. The maximum possible revolution rate is 350 Hz, and the minimum burst time is 7.6 µs. Three sets of fast disk choppers are employed as a pulse shaping chopper (No. 1 in Table 4), a monochromator (No. 2 in Table 4), and a RRM frame overlap choppers (No. 3 in Table 4), and special care has been taken ensure that it works properly under multi-E_i measurement conditions [59]. These are located to satisfy the so-called RRM condition [60]. AMATERAS also has two sets of slow disk choppers that operate at 12.5 Hz or 25 Hz. These choppers have variable opening windows, which can be set from 0° to 175° and are used for frame over suppression and band definition. The detector bank can accommodate 448 ^{3}He 1D PSDs, 3 m in length (2.91 m in effective length), and covers a 0.67π scattering solid angle. We have installed 60% of the required number of detectors currently. The beam-transport of AMATERAS is designed to minimize background from fast neutrons and gamma rays from the source by utilizing curved guide geometry along the horizontal and to maximize the flux at the sample position by employing an elliptical-based geometry determined by series of studies on possible geometries [61–63]. Super mirrors with mirror indices $m = 3.0$ and 3.8 are mainly used. The AMATERAS shielding consists mainly of concrete and iron. Voids along the beamline are filled with concrete, iron, and polyethylene. The housing that accommodates the scattering chamber was constructed using precast concrete panels lined with 35-mm-thick B_4C (25 wt %) mortar plates. Recently, we have started placing B_4C plates on the floor under the scattering chamber because they can reduce the time-independent background.

Table 4. Specifications of AMATERAS (see [9] for details).

Beamline	BL14
Moderator	Coupled hydrogen moderator
Flight path length	$L_{moderator-sample} = 30$ m $L_{sample-detector} = 4$ m
Incident energy (designed)	1–80 meV
Energy resolution	$\Delta E / E_i \geq 1\%$ @ $E_i = 20$ meV
Q resolution (designed)	$2\% > \Delta Q / k_i > 0.2\%$
Detector coverage (scattering angle)	Horizontal: $-40° - +140°$ (final state) $+5° - +110°$ (current state) Vertical: $-16° - +23°$
Detector system	^{3}He 1D-PSD [$\phi = 25.4$ mm, $L = 2910$ mm (effective length)] 448 tubes after full installation (266 tubes are currently installed) Radius 350 mm
Fast disk choppers	Revolution rate ≤ 350 Hz No. 1 (Pulse shaper) $L_{moderator-chopper}$ 7.1 m No. 2 (Monochromator) $L_{moderator-chopper}$ 28.4 m No. 3 (RRM frame overlap) $L_{moderator-chopper}$ 14.2 m Radius 350 mm
Slow disk choppers	Revolution Rate 12.5 Hz or 25 Hz Opening angle 0°–175° (variable) No. 1 $L_{moderator-chopper}$ 9 m No. 2 $L_{moderator-chopper}$ 13.7 m
Beam transport	$m = 3$ and 3.8 supermirror
Beam size at the sample position	Optimal dimensions for beam are 20 mm width × 10 mm height The maximum dimensions are 30 mm width × 50 mm height

On-beam commissioning of the spectrometer started in May 2009, and it was opened to users in December 2009. As mentioned previously, AMATERAS can cover many fields of investigation. Over its seven years of operation, more than 100 proposals have been carried out. Half of them are related to magnetism and strongly correlated electron systems. Studies on magnon–phonon coupled dynamics in multiferroic systems, spin dynamics in frustrated magnets and novel quantum spin systems, crystal field excitations and the underlying electronic properties have been carried out. A quarter of the studies at AMATERAS are related to the field of liquids, glasses, and other non-crystalline systems. Bosonic excitations, fluctuations related to super ion conducting, details of diffusive process in molecules in liquids are being studied. Studies on polymers and biomaterials are also being conducted using AMATERAS. Industrial applications, such as studies on the dynamics of atoms and molecules in rubber materials and on lattice dynamics in plating materials, are also being carried out. One should note that multi-E_i opportunity at this spectrometer has triggered attempts to perform novel data analysis, as described later, such as the mode-distribution analysis, which is a new approach to analyzing quasielastic neutron scattering data without any model assumptions [64].

2.4. Polarized Neutron Spectrometer, POLANO

POLANO is the younger generation of spectrometers operating at MLF [17–19]. As mentioned above, three direct geometry chopper spectrometers are now part of the user program. They cover rather wide energy and momentum spaces for the investigation of the dynamical structures on internal degrees of freedom of materials. POLANO was designed in a way similar to the other three instruments, except it can be used for polarization analysis. In light of recent discoveries in material science, many of the observed complex phenomena are largely caused by entangled physical degrees of freedom (spins, charges, orbitals, and even lattice vibration). A unique, effective, and direct way to observe these physical quantities separately is via polarization analysis.

POLANO is a collaborative project between KEK and Tohoku University under the aegis of KEK Inter-University Research Program, and it commenced in 2009. The construction of the instrument started in 2012, was almost complete in 2016, and radiological assessment for acceptance of the neutron beam was conducted successfully. Consequently, it was just approved as a proper neutron beamline, and its commissioning has just commenced.

Our principal concept is to achieve polarization analysis of inelastic scattering at energies beyond the range of reactor-based neutron sources. A schematic view of the POLANO spectrometer is depicted in Figure 8, and its fundamental specifications are listed in Table 5. POLANO is located at BL23 in MLF, viewing the decoupled hydrogen moderator. The geometrical parameters of the spectrometer are as follows: distance from moderator to sample L_1, distance from sample to detector L_2, and distance from Fermi chopper to sample L_3 are 17.5 m, 2.0 m and 1.85 m, respectively. To ensure the presence of additional space after the Fermi chopper, a rather long L_3 was adopted, without losing much resolution. Energy resolution of $\Delta E/E_i = 3$–5% and momentum resolution of $\Delta Q/k_i = 1$–2% were achieved. These values are sufficient for viewing most magnetic excitations and to observe the incommensurate structure in cuprate high-T_c superconductors, separately.

The pulse width of the decoupled moderator can be determined as $\Delta t_m = a/(E_i)^{1/2}$, where $a = 2.5$. After the moderator, the neutrons are transported by supermirror guide tubes with $m = 4$, optimized for neutron energy of 110–120 meV. An elliptical guide, which is optimized section-wise (section length = 50 cm), can yield a neutron flux of 3.9×10^5 neutrons/(s·meV·cm^2·MW), which is almost comparable to the coupled moderator beamline at $E \sim 100$ meV. The focusing guide tubes affect beam divergence, and therefore, beam profile can be estimated. An SEOP ^{3}He filter cell will be placed at $L = 16.4$ m from the moderator. The ^{3}He filter has a beam width of 8 cm at the bottom and 4 cm at half height. In the early stages, ^{3}He cells with a diameter of 5 cm will be used for neutron spin polarization. A bending mirror with $m = 5.5$ can be used as the spin analyzer placed after sample position in a large vacuum chamber. Relatively high energies of up to 42 meV can be made available for polarization analysis. The vacuum chamber is composed of three sections and is designed to be detachable. The first chamber is the sample chamber, in which the sample is placed and the sample environments are set. The second is a connecting chamber that connects the sample chamber and the scattering chamber. Both the sample and the scattering chambers are sealed by thin ($t = 1.5/2.0$ mm) aluminum windows, and the chambers are completely isolated from air. Certain magnetic devices such as the spin flipper (positioned after the sample) will be installed in this section. The third is the scattering chamber, wherein the suite of analyzer mirror and detectors are placed. In addition, B_4C vanes and liners are installed in the chamber.

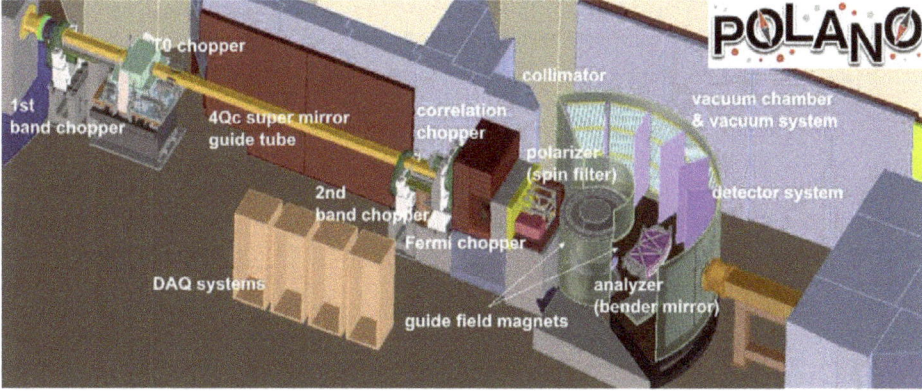

Figure 8. Schematic POLANO beamline and instruments. Neutron beam coming from left-hand side is injected into vacuum chamber (sample) [38].

Table 5. Specifications of POLANO.

Beamline	BL23
Moderator	Decoupled hydrogen moderator
Flight path length	$L_{\text{moderator-sample}} = 17.5$ m $L_{\text{sample-detector}} = 2.0$ m $L_{\text{monochromating chopper-sample}} = 1.85$ m
Incident energy	(unpolarized) 1–500 meV (polarized) 1–100 meV (designed)
Energy resolution	$\Delta E/E_i \geq 4\%$ @ elastic
Q resolution (designed)	1 to 2% $= \Delta Q/k_i$
Detector coverage (scattering angle)	Horizontal: $-20°$ to $+120°$ Vertical: $-8°$ to $+8°$
Detector system	^{3}He 1D-PSD [$\phi = 19$ mm, $L = 600$ mm (effective length)]
Beam transport	$m = 4.0$ supermirror
Beam size at the sample position	Optimal dimensions for beam are 20 mm width $\times$ 20 mm height The maximum dimensions are 50 mm width $\times$ 50 mm height

In POLANO, we primarily target the research field of so-called *strongly correlated electron systems*, where both itinerant and localized electrons play an important role in recent diverse and complicated (interplay between spin, charge, orbital and lattice) science. As is well known, the neutron polarization analysis technique is essential to separately obtain contributing cross-sections. The generalized polarization cross-section can be written with initial and final polarization P_i, P_f as follows,

$$P_f \frac{d^2\sigma}{d\Omega dE_f} = P_i(N^* \cdot N) + (P_i \cdot M_\perp^*)M_\perp + (P_i \cdot M_\perp)M_\perp^* - P_i(M_\perp^* \cdot M_\perp) + iN(P_i \times M_\perp^*)$$

$$-iN^*(P_i \times M_\perp) + NM_\perp^* + N^*M_\perp - i(M_\perp^* \times M_\perp) \tag{1}$$

Here, N and M are the nuclear and the magnetic cross-sections, respectively. In the equation, the first term represents the nuclear cross-section, the following three terms are related to magnetic scattering, and the terms that follow represent nuclear magnetic interference and magnetic chirality.

Thus, we can extract some part of the cross-section by measuring the combination of spin polarimetry up and down, namely, I^{++}, I^{--}, I^{+-}, and I^{-+}.

We started spin polarization experiments up to $E_f < 50$ meV as the first step. Since the energy scale in heavy fermion or 4f electron systems is relatively low, it is one of themes carried out in the beginning. The search for direct evidence of orbital waves and multiferroic materials will also be the target to be explored. With the development of the wide-angle SEOP device, the energy range will be expanded to up to approximately 120 meV, entering a new paradigm for neutron polarization analysis. The reactor-based instruments no longer reach this high energy level, and the scheduled accelerator-based instruments are designed for lower energy range (cold neutrons). In this energy region, many phonon frequencies (Debye frequencies) are within this energy range in the case of solid state materials. Even high-T_c superconductivity will be objective. Hydrogen science can also be the target of research with neutron polarization analysis because the dynamical motion of hydrogen in most metal hydrides can be roughly described in the energy range of 70–150 meV energy as local vibrational modes.

We expect to realize neutron polarization at much higher energy as the final step. This needs revolutionary techniques to polarize sub-eV neutron spins, such as dynamical neutron polarization (DNP) or high polarization SEOP/MEOP. Once we obtain these high-energy neutron polarization techniques, research on itinerant electron systems and spin-charge separation in strongly correlated electron systems such as cuprates will progress remarkably.

2.5. Biomolecular Dynamics Spectrometer, DNA

DNA, a time-of-flight (TOF) near-backscattering spectrometer (n-BSS), is a unique instrument among spallation pulsed neutron facilities over the world in terms of n-BSS equipped with a high-speed pulse-shaping disk-chopper [12]. Neutron beams from the coupled moderator, which provides the most intense but broadest pulse among all three moderators in MLF, is handled flexibly in terms of pulse width by this chopper, while maintaining its intensity and symmetry in the TOF spectrum. Si crystal analyzers back-coated with the neutron absorber reduce unfavorable background scattering of the instrument dramatically, achieving signal-to-noise ratios of ~10^5. Those factors are important for extending the application fields to dynamical behaviors of atoms and spins in bio-molecules, soft-materials, and strongly-correlated electron systems at the nanosecond timescale or in the micro-eV energy region.

Two visual representations (a three-dimensional (3D) view and an inside view of the scattering vessel) of DNA are shown in Figure 9a,b. The characteristic parameters are listed in Table 6. The Si-111 crystal analyzers, the first key device for DNA, installed in the vacuum vessel cover scattering angles ranging from $-30°$ degrees to $+150°$ in the horizontal plane and from $+21°$ to $-14°$ in the vertical plane (Figure 9b). The measurable momentum range is $Q = 0.08–1.86$ Å^{-1}. The analyzer spherical surface is divided into the upper and the lower parts centered at different positions 102 mm upward and downward of the scattering center (sample position), respectively. Neutrons scattered by the sample are energy-analyzed under the near-back-scattering condition with Bragg angle θ_B ~$87.5°$ and then countered by position-sensitive ^{3}He gas detectors (PSDs) arranged on the circumference of the same scattering center and shifted upward and downward (Figure 9b). These arrangements of the analyzer units and the detectors allow us to angle-resolve scattered neutrons in the horizontal and the vertical directions two-dimensionally, and therefore, this spectrometer has the potential for use in experiments involving single crystalline samples.

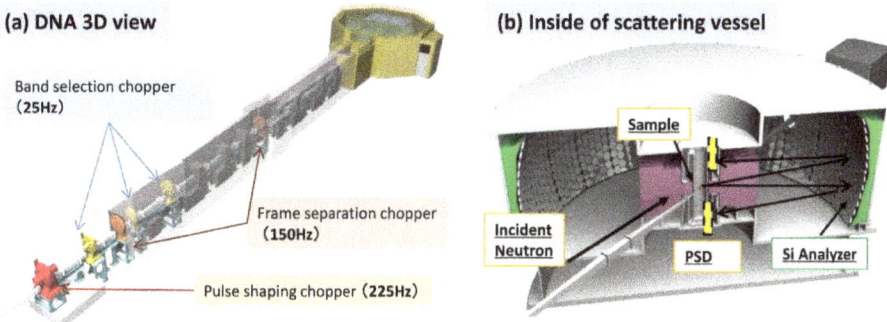

Figure 9. DNA three-dimensional view (**a**) and inside of the scattering vessel (**b**).

Table 6. Specifications of DNA.

Beamline	BL02
Moderator	Coupled Hydrogen Moderator
$L_{\text{moderator-sample}}$	42 m
$L_{\text{sample-analyzer}}$	~2.3 m
$L_{\text{analyzer-detector}}$	~2.0 m
Pulse sharpening chopper (PS-chopper)	At ~7.5 m from the moderator Max speed: 300 Hz (designed value) (Present maximum speed: 225 Hz) 4 slits on one disk
Crystal analyzer	Crystal and reflection index Si(111) Si(311) in test Bragg angle of analyzers ~87.5°
Energy resolution	~2.4 µeV: Si-111 with 10mm Slit @225Hz of PS-chopper ~3.5 µeV: Si-111 with 30mm Slit @225Hz of PS-chopper ~14 µeV: Si-111 without PS-chopper ~12 µeV: Si-311 with 10mm Slit @225Hz of PS-chopper
Momentum range	$0.08 < Q < 1.86$ Å^{-1}: Si-111 $1.0 < Q < 3.80$ Å^{-1}: Si-311 (in plan)
Scan energy range	Si-111 $-40 < E/\mu eV < 100$: Single pulse scan around E_f $-400 < E/\mu eV < 600$: Multi pulse scan around E_f $-500 < E/\mu eV < 1500$: without PS-chopper in second frame Si(311) $-150 < E/\mu eV < 300$: Single pulse scan around E_f (the specifications by the end of March 2017)

The pulse-shaping chopper, the second key device in DNA, is located relatively upstream in the neutron guide at 7.75 m from the moderator because the time for slit opening at this position becomes the origin of TOF analysis. The chopper disk has four slits with two types of different slit sizes (3 with 30 mm slit and 1 with 10 mm slit), which allows us to adjust the energy resolution and rotational speed. The measurable energy resolutions are $\Delta E = 2.4$ and 3.6 µeV with the 10-mm slit and the 30-mm slit, respectively. Another merit of the pulse-shaping chopper is that allows for the realization of a scanning method over a wide energy-transfer range while maintaining high energy resolution by changing the phase of slit opening relative to pulse trigger.

The DNA spectrometer is helping us to advance a plan to expand measurable momentum transfer Q range. For that purpose, 10 analyzer-mirror sets for determining the Bragg reflection of Si-311 were fabricated. Those ten Si-311 analyzer mirror sets will be installed in the vacuum vessel to cover scattering angles ranging from −30° to −150° by October 2018. The expected measurable momentum transfer range will be $Q = 1.0–3.8$ Å^{-1}, and the expected energy resolution will be $\Delta E = 12$ μeV with the 10-mm slit. With this advancement, measurements at high-Q with high energy resolution will become possible.

As another improvement plan, we are proceeding with the installation of three sets of diffraction detectors at a high scattering angle. Three sets of diffraction detector banks will be installed at scattering angles of around $2\theta = +159°$, $+153°$, and $−159°$, respectively, for the following purposes: to evaluate the energy distribution of an incident neutron beam from a measurement of a standard vanadium sample and to obtain the long-d diffraction pattern from powder samples with high resolution $\Delta d/d$ < 0.5%. Such a diffraction measurement would also contribute to the evaluation of accuracy of inelastic scattering measurements.

Because of its expanded measurable energy range along with high energy resolution and low instrumental background, the DNA spectrometer is in a new category of back-scattering spectrometers with micro-eV energy resolution, different from conventional back-scattering spectrometers. Accordingly, the target sciences are expanded not only for quasi-elastic neutron scattering but also for inelastic neutron scattering with very high energy resolution. It is expected to increase the number of experimental proposals from the research field of strongly-correlated electron systems, such as low-energy excitation at low temperatures.

2.6. Village of Neutron Resonance Spin Echo Spectrometers, VIN ROSE

Two NSE spectrometers with resonance neutron spin flippers, that is, a NRSE instrument and a MIEZE instrument, have been installed by Kyoto University and KEK jointly at BL06 for viewing the coupled moderator. NSE is an essential spectroscopic technique, which has achieved the highest energy resolution with neutrons as the probe [65]. NSE with RSFs is a rather new approach [66]. In addition, the combination of NSE and a pulsed neutron source makes possible to scan a wide spatiotemporal space very efficiently. The beamline has been named VIN ROSE (VIllage of Neutron ResOnance Spin Echo spectrometers), and it will spawn many new field of spectroscopic methods [20]. Figure 10 shows a schematic top view of BL06. The main specifications of VIN ROSE are listed in Table 7.

NRSE is suitable for studying the slow dynamics of soft condensed matter with high energy resolution, while MIEZE offers the advantage of flexible sample environments, especially for magnetic field applications, at the sample position. The designed dynamic range of MIEZE is $0.2 < Q < 3.5$ Å^{-1} and $0.001 < t < 2$ ns, and that of NRSE is $0.02 < Q < 0.65$ Å^{-1} and $0.1 < t < 100$ ns, respectively. So far, the most successful application of NSE was to polymer systems, as reviewed in Ref. [67], and these systems are the target of NRSE. Owing to the abovementioned characteristics, MIEZE is expected to open new fields of study such as spin dynamics in strongly correlated systems [68–70].

The project was started in Japanese fiscal year (JFY) 2011, and the first neutron beam was accepted in JFY2014. After that, the performance of the installed neutron guides was tested using several methods. In the case of NRSE, the neutron flux observed at the guide-end position agreed well (95%) with the calculated value (6.9×10^8 neutrons/(cm^2·s) at 1MW). By contrast, the agreement was 56% for MIEZE (2.7×10^8 neutrons/(cm^2·s) at 1 MW) [71]. The first MIEZE signal was observed at the end of JFY2014. The combination of TOF and MIEZE (TOF-MIEZE) is a novel approach [72], and its detailed verification is in process. In the case of continuous neutron sources, the geometric restriction of MIEZE is very tight, that is, a very accurate setting for the distances among the source, flippers, and the detector is required. However, this requirement can be relaxed by using the TOF-MIEZE method, which was investigated both theoretically and experimentally by using BL06 [71]. Figure 11 shows the measured 200-kHz TOF-MIEZE signal and the corresponding power spectrum.

MIEZE will be open to user programs in the autumn of 2017. The program with NRSE will be delayed by a year.

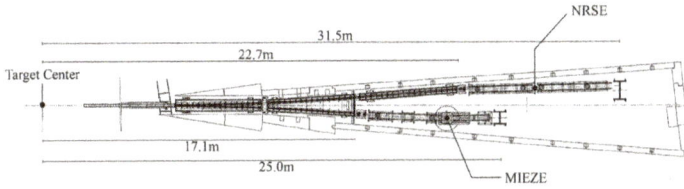

Figure 10. Schematic top view of the Modulated Intensity with Zero Effort (MIEZE) and Neutron Resonance Spin Echo (NRSE) spectrometers at BL06.

Table 7. Specifications of VIN ROSE.

Beamline	BL06
Moderator	Coupled hydrogen moderator
Measured neutron flux and peak wavelength of guide exit	MIEZE: 2.7×10^8 n/cm²/s/MW, 4.8 Å NRSE: 6.9×10^8 n/cm²/s/MW, 5.5 Å
Band disk choppers	Radius 200 mm Revolution Rate 25 or 12.5 (or 8.33) Hz Opening angle 126° (MIEZE) Opening angle 162° (NRSE) Moderator-chopper 12.1 m
Ellipsoid focusing mirror (NRSE)	Semi-major axis: 1.25 m Semi-minor axis: 65.4 mm Sample size(NRSE) < 5 mm × 5 mm
Wavelength, Q range, Fourier time	MIEZE $3 < \lambda < 13$ Å, $0.2 < Q < 3.5$ Å^{-1}, 1 ps $< t < 2$ ns [*1] NRSE $5 < \lambda < 20$ Å, $0.02 < Q < 0.65$ Å^{-1}, 0.1 ns $< t < 0.1$ µs [*2]

[*1] The maximum Fourier time depends on sample size. In the case of a thin film with reflectometry, the maximum Fourier time can be extended. [*2] The maximum Fourier time is a first target value. It depends on the performance of ellipsoid-focusing supermirror.

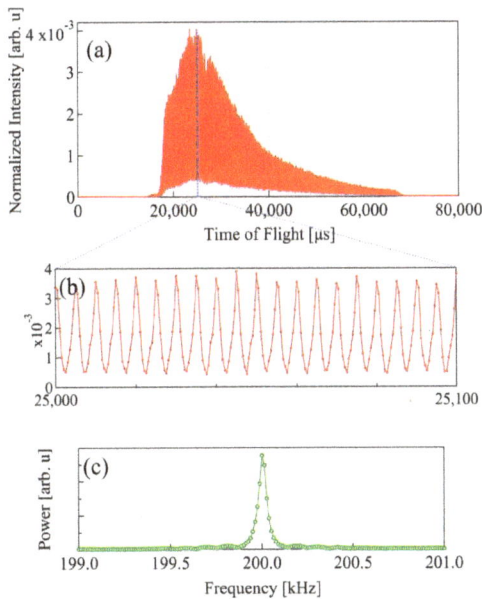

Figure 11. (**a**) Measured time-of-flight (TOF)-MIEZE signal with an effective frequency of 200 kHz; (**b**) TOF-MIEZE signal between 25 and 25.1 ms, and (**c**) power spectrum of TOF-MIEZE signal by Fourier transformation.

3. Crystal Structure Group Instruments

The MLF Crystal Structure Group consists of seven instruments for powder and single-crystal diffraction. SuperHRPD, SPICA, and iMATERIA are powder diffractometers with detectors having wide angular coverage. SuperHRPD has a very long flight path length to achieve its high resolution. SPICA adopts an elliptical neutron guide to achieve high incident flux. iMATERIA optimizes the flight path length and the neutron guide to cover a wide lattice spacing (d)-range along with high incident flux. PLANET and TAKUMI are diffractometers focusing on small gauge volumes for special purposes. PLANET has a relatively high incident flux and wide d-range for high-pressure studies, while TAKUMI has a relatively high resolution optimized for engineering studies. iBIX and SENJU are single-crystal diffractometers. iBIX is an atmospheric sample-space type instrument, while SENJU is a vacuum sample-space type instrument. Figure 12 shows the covering d-ranges and resolutions ($\Delta d/d$) covered by the diffractometers. The d-range of iMATERIA is the widest, and the $\Delta d/d$ value of SuperHRPD is the smallest.

The notation of d is mainly used in this section because most of the instruments in this group deal with crystal analysis. For some instruments that also deal with liquid, amorphous and small angle scattering, the notations of both d and Q values are used.

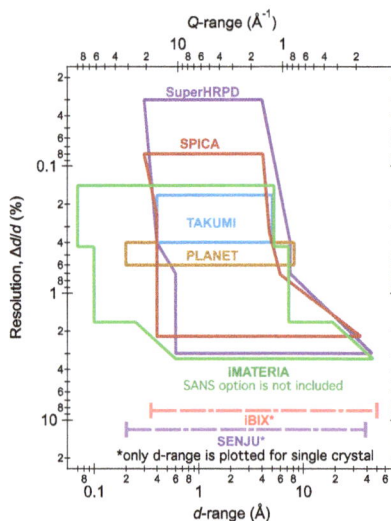

Figure 12. *d*-ranges and peak resolutions of diffractometers in diffraction sub-group. Note that only the covering *d*-range is plotted for single-crystal diffractometer of iBIX or SENJU.

3.1. Super High-Resolution Powder Diffractometer, SuperHRPD

Super High-Resolution Powder Diffractometer, SuperHRPD [21], is located at about 100 m from the thin side of a decoupled poisoned moderator at MLF in J-PARC. SuperHRPD consists of a vacuum sample chamber with capacity of about 1 m^3 and gas-filled scattering banks around it. To cover this large detector solid angle, about 1300 1D ^{3}He PSDs have been installed in the backward bank, 90° scattering bank, and low-angle scattering bank (Figure 13a). Table 8 shows the instrumental specifications of SuperHRPD. SuperHRPD is expected to (1) detect tiny distortions never detected before in the field of strongly correlated systems, multiferroics, magnetic materials; (2) facilitate high-precision structural analysis of energy materials; (3) develop structure science of hybrid materials, supermolecules, and pharmaceuticals. Various achievements have been made in many fields by the user program and the instrumental group project that commenced in December 2008 [73–80].

The instruments are being improved continually, and a new detector system using PSDs measuring 8 mm in diameter was installed at all backward banks in 2014. The overall resolution of the standard powder sample improved from $\Delta d/d = 0.1\%$ to 0.08% [81]. By using a limited detector area for the horizontal part corresponding to the highest scattering angle, $\Delta d/d$ of less than 0.04% was achieved. Figure 13b shows a peak-fitting result of this area obtained by using the asymmetric pseudo-Voigt function. $\Delta d/d = 0.0365(1)\%$ is obtained, which is equivalent to the value evaluated using a single crystal of silicon in June, 2008. This result shows the possibility of carrying out experiments with very high resolution when using a powder sample by using the limited detector area.

Table 8. Specifications of SuperHRPD.

Moderator	Poisoned Decoupled Hydrogen Moderator
Primary flight path $L_{\text{moderator-sample}}$	94.2 m
Curved guide	31.245 m ($m = 3$, $r = 5$ km)
Straight guide	51.4 m ($m = 3$)
Position for disk choppers	7.1 m (single), 12.75 m (double)
Backward bank	
2θ	$150° \leq 2\theta \leq 175°$
$L_{\text{sample-detector}}$	2.0–2.3 m
d-range	0.3–4.0 Å
Resolution $\Delta d/d$	0.03–0.1%
90° scattering bank	
2θ	$60° \leq 2\theta \leq 125°$
$L_{\text{sample-detector}}$	2.0–2.3 m
d-range	0.4–7.5 Å
Resolution $\Delta d/d$	0.4–0.7%
Low angle scattering bank	
2θ	$10° \leq 2\theta \leq 40°$
$L_{\text{sample-detector}}$	2.0–4.5 m
d-range	0.6–45 Å
Resolution $\Delta d/d$	0.7–3.0%
Ancillary equipment and sample environment	Auto sample changer (RT, 10 samples) 4 K-type closed cycle refrigerator (4–300 K) Top-loading refrigerator (10–300 K) High temperature furnace (950 °C)

m: *m*-values of supermirror guide; *r*: curvature radius of curved guide; RT: room temperature.

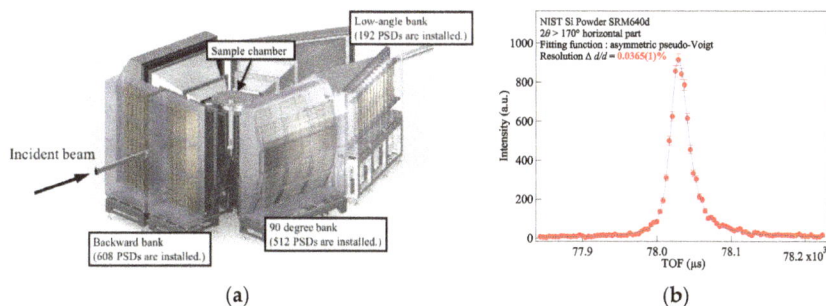

(a)

(b)

Figure 13. (**a**) 3D-image of SuperHRPD diffractometer. About 1300 PSD detectors are installed, where those with better resolution are in the backward detector bank; (**b**) Diffraction peak of the horizontal part of the area with the highest scattering angle having $\Delta d/d = 0.0365(1)\%$.

3.2. Special Environment Powder Diffractometer, SPICA

SPICA (Figure 14), a special environment powder neutron diffractometer, was built at BL09. This is the first instrument dedicated to the study of next-generation batteries, and it is optimized for *operando* measurement to clarify structural changes at the atomic level in the materials used in battery devices. To observe real-time structural changes in materials used in practical devices, a diffractometer with a good balance between intensity and resolution is needed. Therefore, SPICA was optimized to have the sample position of 52 m from the thin side of the decoupled poisoned moderator for achieving high resolution and to adopt an elliptical supermirror guide tube to enhance neutrons at the sample position. As a result, the highest $\Delta d/d$ of 0.08% is achieved at the back-scattering bank, and the neutron intensity increased to four times higher than that of the tapered-straight guide [22]. The specifications of SPICA are summarized in Table 9.

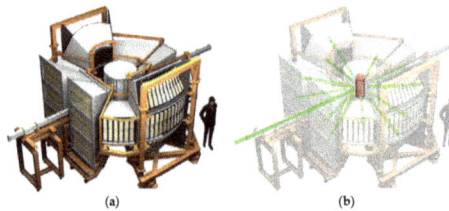

(a)   (b)

Figure 14. (a) Overview of SPICA instrument. The horizontal detector crates are arranged on a cylindrical locus from $2\theta = 5°$ to $175°$. (b) Illustration of *operando* measurements. An 18650-type cell is placed at the center of the sample chamber. The neutron incident beam enters the sample chamber, where it is diffracted by the electrodes, and subsequently reaches the detectors.

Table 9. Specifications of SPICA.

Moderator	Poisoned-Decoupled Hydrogen Moderator
Primary flight path $L_{\text{moderator-sample}}$	52.0 m
Supermirror guide	39.4 m ($m = 3$–6, $r = 5$ km)
Position for disk choppers	7.25, 12.8, 18 m (Single)
Backward bank	
2θ	$149° \leq 2\theta \leq 172°$
$L_{\text{sample-detector}}$	2.0–2.3 m
d-range	0.3–4.1 Å *
Resolution $\Delta d/d$	0.08–0.15%
High scattering bank	
2θ	$116° \leq 2\theta \leq 139°$
$L_{\text{sample-detector}}$	2.0 m
d-range	0.4–4.7 Å *
Resolution $\Delta d/d$	0.24–0.33%
90° scattering bank	
2θ	$66° \leq 2\theta \leq 110°$
$L_{\text{sample-detector}}$	2.0 m
d-range	0.4–6.0 Å *
Resolution $\Delta d/d$	0.37–0.71%
Low angle scattering bank	
2θ	$14° \leq 2\theta \leq 58°$
$L_{\text{sample-detector}}$	2.0 m
d-range	0.6–35 Å *
Resolution $\Delta d/d$	0.8–2.2%
Ancillary equipment and sample environment	Auto sample changer (RT, 40 samples) Top-loading refrigerator (10–300 K) Top-loading cryo-furnace (20–800 K)

m: *m*-values of supermirror guide; *r*: curvature radius of curved guide, *d-range is obtained under the 25/3 Hz mode. The frequency can be tuned until 25/5 Hz.

For real-time measurement of a practical device under an operative condition, SPICA provides space for samples with diameters as large as 2 m diameter, which makes it a very favorable sample-handling environment. Because the *operando* measurements are mostly carried out in air, coarse collimators with B_4C-resin-coated aluminum blades are installed in the air-scattering chambers to reduce air scattering. In the previous simulation study, the loss of neutrons in the air-scattering chamber was approximately 6–11% over the wavelength range of 0.2–6.0 Å [82].

Figure 15 shows a typical *operando* measurement of a commercial Li-ion battery. Structural changes in the active materials, which depend on the lithium content, can be observed clearly. The lattice parameters of the anode and the cathode materials were extracted from the diffraction patterns as a function of the lithium content. An automatic data analysis procedure was developed, as shown schematically in Figure 16. Z-Rietveld, a Rietveld refinement software developed in J-PARC [83], can handle large numbers of diffraction data provided by the *operando* observation and refine the structures of both electrodes sequentially. The full-scale usage of SPICA has just begun. The first battery study on the structure dependence of the charge–discharge rate was published in 2016 [84]. SPICA will be used to carry out experiments on material science and observe structural changes in practical devices under an operative condition. Finally, it will be used to understand what happens in a device during its operation and for the development of next-generation batteries.

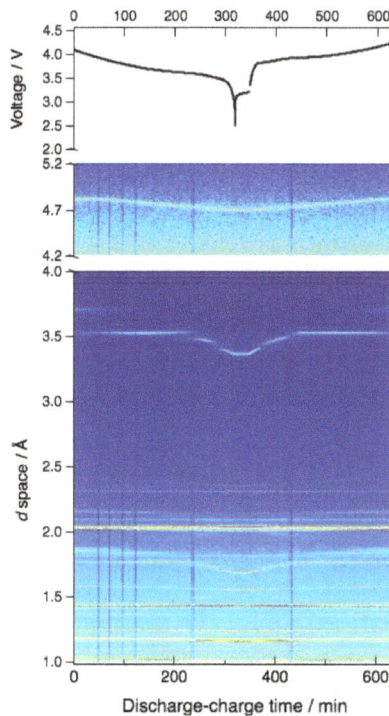

Figure 15. Changes in relationship between discharge/charge time and diffraction profiles of materials in 18,650-type lithium battery.

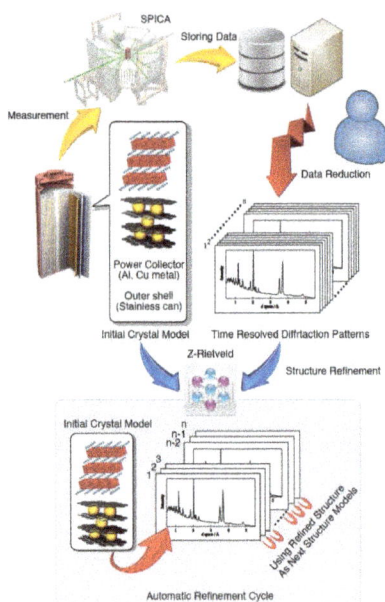

Figure 16. Schematic diagram of flowchart for automatic refinement cycle.

3.3. IBARAKI Materials Design Diffractometer (Versatile Neutron Diffractometer), iMATERIA

Ibaraki prefecture, where the J-PARC site is located, has decided to build a versatile neutron diffractometer (IBARAKI Materials Design Diffractometer (iMATERIA) (see Figure 17) [23]) to promote industrial applications by using the neutron beam in J-PARC. iMATERIA is planned to be a high-throughput diffractometer that could be used by materials engineers and scientists for their materials development work, like the chemical analytical instruments. The applications of neutron diffraction in materials science are as follows: (1) structural analyses of newly developed materials, (2) determination of the correlation between structures and properties (functions), and (3) determination of the relationship between structural changes and improvements of functions, especially for practical materials. To achieve those purposes, a diffractometer with super high resolution is not required. The balance among intermediate resolution around $\Delta d/d = 0.15\%$, high intensity, and wide d coverage is more important.

This diffractometer is designed to look at a decoupled-poisoned supercritical hydrogen moderator (36 mm, off-centered) (BL20) and to have the incident flight path (L_1) of 26.5 m with three wavelength selection disk-choppers and straight neutron guides having a total length of 14.0 m. The instrumental specifications are listed in Table 10. There are four detector banks, including a low-angle and small-angle scattering detector bank. The angular coverage of each detector bank is also shown in Table 10. When the rotation speeds of the disk-choppers are the same with a pulse repetition rate of 25 Hz (SF mode), the diffractometer covers $0.18 < d$ (Å) < 2.5 with $\Delta d/d = 0.15\%$ and covers 2.5 $< d$ (Å) < 800 at three detector banks of 90°, low angle, and small angle with gradually changing resolution. When the speed of the wavelength selection disk-choppers is reduced to 12.5 Hz (DF mode), we can access a wider d-range, $0.18 < d$ (Å) < 5 with $\Delta d/d = 0.15\%$, and $5 < d$ (Å) < 800 with gradually changing resolution with doubled measurement time from the SF mode. All four banks, including the small-angle bank, are user operational. It takes about 5 min (DF mode) to obtain Rietveld-quality data in the high-resolution bank with 500 kW of beam power for about 1 g of standard oxide samples. The automatic sample changer is the most important sample environment for high-throughput experiments. Our automatic sample changer [85] consists of sample storage, elevating system of two lines, two sets of pre-vacuum chambers, and a sample-sorting system. We can handle more than 600 samples continuously at room temperature without breaking the vacuum of the sample chamber.

Figure 17. IBARAKI Materials Design Diffractometer, iMATERIA without detector for each bank and instrument shielding. The high-resolution bank, special-environment bank (90° bank), and low-angle bank, can be seen from right to left. The small-angle detector bank, which is not shown in the picture, is situated in the low-angle vacuum chamber (left hand side of the picture).

Table 10. Specifications of iMATERIA. The d-range (Q-range) of each bank is the maximum value in the two-measurement mode.

High	2θ	$150° \leq 2\theta \leq 175°$
Resolution	$L_{\text{sample-detector}}$	2.0–2.3 m
Bank	d-range	$0.09 \leq d\,(\text{Å}) \leq 5.0$
Special	2θ	$80° \leq 2\theta \leq 100°$
Environment	$L_{\text{sample-detector}}$	1.5 m
Bank	d-range	$0.127 \leq d\,(\text{Å}) \leq 7.2$
Low	2θ	$10° \leq 2\theta \leq 40°$
Angle	L_2	1.2–4.5 m
Bank	d-range	$0.37 \leq d(\text{Å}) \leq 58$
Small	2θ	$0.7° \leq 2\theta \leq 5°$
Angle	$L_{\text{sample-detector}}$	4.5 m
Bank	Q-range	$0.007 \leq q(\text{Å}^{-1}) \leq 0.6$

3.4. High-Pressure Neutron Diffractometer, PLANET

PLANET is the beamline dedicated to high-pressure experiments [24]. By using various high-pressure devices, powder diffraction data can be obtained over wide pressure and temperature ranges of 0–20 GPa and 4–2000 K, respectively. The beamline is equipped with narrow incident slits and radial receiving collimators, which enable us to obtain very clean patterns without contamination of Bragg peaks by the materials surrounding the sample, such as a heater and a sample container. This contributes to precise structure determination of crystals and liquid/amorphous materials under high-PT conditions.

Figure 18 shows a cutaway view of the beamline. The characteristic parameters are listed in Table 11. The beamline is designed by adding the components specific to high-pressure experiments to a conventional powder diffractometer setup. Neutron beam originating from the moderator are truncated with an iron collimator spatially and choppers energetically, and useful neutrons are transferred to the experimental hatch by using a focusing mirror guide. They are truncated again by the second slits in front of the experimental hatch and are adjusted into smaller size than the sample by the third slits. The neutrons are introduced to the sample under high pressure and those diffracted toward 90° are detected using the 90° detector banks consisting of 320 PSDs. The detailed design and specifications are described in [24]. The high-pressure and high-temperature condition is generated using the six-axis multi-anvil press ATSUHIME designed for TOF measurements [86]. In this system, the sample is compressed isotropically using six anvils made of tungsten carbide. The mutual positions of the six anvils are controlled precisely to a precision of a few microns to ensure isotropic compression. The large sample space makes it possible to assemble a hydrogen source in the high-pressure cell, which can be used to study hydrides stable only under high-PT conditions [87]. The high-pressure and low-temperature condition is generated using the Mito system [88]. In this system, only the sample and the anvil parts are cooled by circulating liquid nitrogen around them, while the hydraulic oil is maintained at ambient temperature. Therefore, the sample pressure can be changed even at low temperatures. This system is applied to the study of ice under low-T and high-P conditions [89,90]. PLANET is widely used in the fields of geoscience, high-pressure physics, and material science.

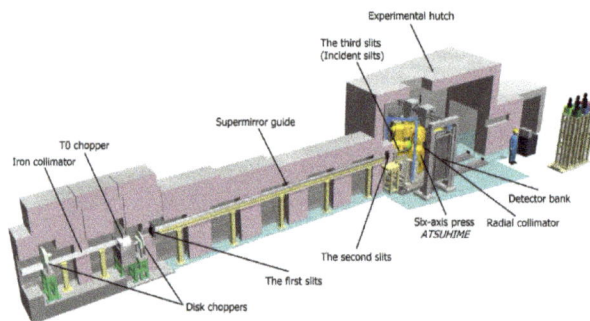

Figure 18. Cutaway view of PLANET beamline. The figure shows the setup for experiments using the six-axis press. Reprinted from [72] with permission from Elsevier.

Table 11. Specifications of PLANET.

Characteristics	Parameters
Moderator	20 K Para-hydrogen (decoupled)
Source-to-sample distance	25 m
Sample-to-detector distance	1.5–1.85 m (depending on the scattering angle)
Detector coverage	90° ± 11.3° (horizontal), 0° ± 34.6° (vertical)
Wavelength	0.3–6.0 Å
Resolution	$\Delta d/d = 0.4$–0.6%
d-spacing range	0.2–4.2 Å (in the single frame mode) 0.2–8.4 Å (in the double frame mode)
Q-value range	1.5–30 Å^{-1} (in the single frame mode) 0.8–30 Å^{-1} (in the double frame mode)
Neutron flux at sample position in 10 mm Ø	5.291×10^7 neutrons cm^{-2} s^{-1} (@1 MW)
Pressure and temperature range	0–16 GPa, RT–2000 K (ATSUHIME) 0–20 GPa, RT (PE-press) 0–10 GPa, 77–473 K (Mito system)

3.5. Engineering Materials Diffractometer, TAKUMI

TAKUMI is a TOF neutron diffractometer dedicated to engineering material sciences. Careful analysis of the Bragg peaks in a neutron diffraction pattern can reveal important structural details of a sample material such as internal stresses, phase conditions, dislocations, and texture. Such information is often crucial in engineering applications, and the ability to carry out either ex-situ or in situ measurements makes neutron diffraction particularly useful in this respect. Detailed information about the design and performance of TAKUMI can be found in [25,26]. TAKUMI is installed at BL19 in MLF, and it has been designed to solve various problems related to engineering materials: (i) internal strain mapping in engineering components; (ii) microstructural evolutions during deformations and/or thermal processes of structural or functional materials; (iii) microstructural evolutions during thermo-mechanical processes of structural or functional materials; and (iv) crystallographic investigations of small regions in engineering materials. For those purposes, TAKUMI is composed of an incident beam slit with adjustable size and distance to the sample position, main detector banks positioned at the scattering angles of ±90°, several radial collimators of various sizes, and a large sample table that can be moved precisely. Figure 19 shows the appearance of TAKUMI.

TAKUMI covers engineering materials such as metals, ceramics, and composites varying in size from small to large. Its specifications are listed in Table 12. The total flight path is designed to cover a wide d-range simultaneously, which is very useful not only for stress mapping but also for various types of in situ observations. The d-range of about 0.25 nm is achieved without sacrificing the high intensity of MLF owing to operation at the same repetition rate as that of J-PARC (25 Hz). This d-range provides more than 15 peaks, including the lowest diffraction index one, for austenitic or ferritic steels. The minimum d and the maximum d can be tuned, or the d-range can be doubled for samples having larger lattice constants. The best peak resolution $\Delta d/d$ is about 0.17%, which facilitates the separation of diffraction peaks from phases having similar crystal structures and to conduct peak broadening analyses, as well as to perform highly accurate strain measurements. Data reduction procedures have been developed to use the event data recording system maximally, as shown in Table 12, and this can be done during or after the experiment. Various types of sample environment devices have been developed and incorporated into TAKUMI, in addition to the common devices in MLF. Some of them were developed together with external users. They are open to all users for their experiments.

As one of the day-one instruments, TAKUMI has been open to the user program since 2009 and has assisted in the production of many publications, typically, papers related to strain mapping [91,92], in situ room-temperature loading [93,94], in situ low-temperature loading [95,96], in situ thermal treatment [97,98], and dislocation characterization [99].

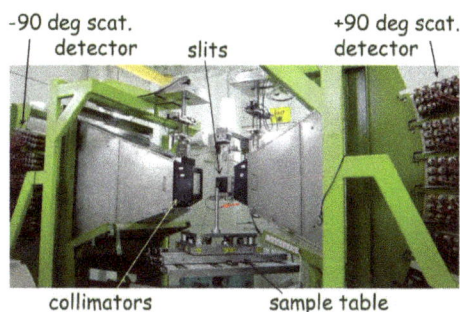

Figure 19. The appearance of TAKUMI.

Table 12. Specification of TAKUMI.

Moderator	Poisoned decoupled hydrogen moderator
d-range	• Standard (25 Hz operation): Δd ~0.25 nm (d_{min} and d_{max} are tunable) • Wide (12.5 Hz operation): Δd ~0.50 nm (d_{max} is about 0.50 nm)
S/N ratio	~10^{-3}
Peak resolution	Tunable ; Low (~0.4 %) Medium (~0.3 %, most cases) High (~0.2 %)
Radial collimators	1 mm, 2mm, 5 mm (a pair of each)
Data acquisition	• Event recording (data reduction with the functions of time, TOF-binning width, and detector range is flexible.) • Data reduction based on the physical conditions (load, strain, temperature, etc.) is under development.
Ancillary equipment and sample environment	a) Large sample table (table size: 700 mm × 700 mm, load capacity: 1 ton) b) BL19 Standard loading machine (50 kN) c) Furnace system for high temp loading (1273 K), added to (b) d) 100 K cooling system for loading experiment (100 K–473 K), added to (b) e) Cryogenic loading machine (6 K–220 K, 50 kN) f) Fatigue machine (60 kN, < 30 Hz) g) High temperature loading machine for small specimen (25 kN, 1273 K) h) Thermo-mechanical simulator (30 kN, 100 mm/s, 1473 K, 30 K/s) i) Multipurpose furnace (1273 K) j) Multipurpose cryostats (4 K, sample size: 100 × 100 × 100 mm³) k) Dilatometer (1273 K) l) Eulerian Cradle m) Gandolfi goniometer

3.6. IBARAKI Biological Crystal Diffractometer, iBIX

Single-crystal neutron diffraction is among the powerful methods to obtain the structural information, including hydrogen atoms. IBARAKI biological crystal diffractometer, iBIX (Figure 20), is a TOF neutron single-crystal diffractometer with high performance in elucidating the hydrogen, protonation, and hydration structures of mainly biological macromolecules in various life processes. To achieve high performance, we developed a new photon-counting 2D detector system using scintillator sheets and wavelength-shifting fiber arrays for the X/Y axes. Since the end of 2008, iBIX has been available to user experiments supported by Ibaraki University [27]. In JFY2012, we succeeded in upgrading the 14 existing detectors and installing 16 new detectors for the diffractometer of iBIX [28]. The total solid angle of the detectors subtended by a sample and the average detector efficiency increased by 2 and 3 times, respectively. The total measurement efficiency of the present diffractometer is one order of magnitude higher than that of the previous one, and the accelerator power has increased. At the end of 2012, it was possible to use iBIX regularly to investigate biological macromolecules in user experiments. The final specifications of the iBIX are given in Table 13. In 2015, the accelerator power of J-PARC reached 400–600 kW. The full data set of biological macromolecules for neutron structure analysis was obtained using iBIX as follows. The maximum unit cell size was set to 110 Å × 110 Å × 70 Å. The average sample volume was set to 2–3 mm^3 and average measurement time to 7–10 days. If the accelerator power were to reach 1 MW, the total measurement time or the sample size will be reduced by half. To improve the quality of the integrated intensity of weak reflections, we developed a profile-fitting method for the peak integration of the data reduction software STARGazer [100]. The integrated intensities and model structure obtained using the profile-fitting method were more accurate than those obtained using the summation integration method, especially in the case of higher-resolution shells [101]. We have already prepared a user manual and a distribution package for the data reduction software, including the profile-fitting component.

In the future, the accelerator power of J-PARC will be improved to 1 MW. iBIX should be available regularly for full dataset measurement with a sample size of 1 mm^3. We will continue to develop the data reduction software and beamline instruments to improve the accuracy of the intensity data obtained from small samples. Furthermore, we will validate whether full data measurement, reduction, and structure analysis of the sample are possible when using a large unit cell (132 Å × 132 Å × 132 Å) in iBIX.

Figure 20. Diffractometer of iBIX.

Table 13. Specifications of iBIX.

Moderator	Coupled
Wavelength of incident neutron	0.7–4.0 Å (1st frame)
	4.0–8.0 Å (2nd frame)
Neutron intensity (@1MW)	0.7×10^6 n/s/mm^2
$L_{moderator-sample}$	40 m
$L_{sample-detector}$	500 mm
Solid angle of detectors	19.5% for 4π
Detector covered region	15.5–168.5°
Detector size	133 mm × 133 mm
Detectors pixel size	0.52 mm × 0.52 mm
No. of detectors	30

3.7. Extreme Environment Single-crystal Neutron Diffractometer, SENJU

Single-crystal neutron diffraction is also a fundamental and powerful tool in materials science because of its availability to determine the arrangement of light elements and magnetic moments in crystalline materials with high accuracy and reliability. Thus, this technique has been an irreplaceable analytical tool for the development of new functional materials such as proton conductors, hydrogen-absorbing materials, and magnets. However, because a large (over 1.0 mm^3 in volume) single-crystal sample is required, the number of experiments has been limited. Recently, we developed a new single-crystal neutron diffractometer called SENJU at MLF to alleviate this limitation.

A schematic view of SENJU is shown in Figure 21a, and its characteristic parameters are listed in Table 14. SENJU has a vacuum sample chamber around which 37 area detectors are arranged cylindrically, as shown in Figure 21a. Owing to proper design of the chamber and collimators, a very low background was achieved; consequently, weak Bragg spots from the small sample crystal were observed clearly. In addition, large amounts of neutrons scattered from the sample were collected efficiently by covering a large solid angle with the area detectors. In addition to being used at the high-intensity neutron beam in J-PARC, these features have also allowed for diffraction measurements of 0.1-mm^3-volume single crystals at SENJU [29]. At SENJU, a two-axis sample-rotation system for temperatures below 10 K was also developed using piezo-rotators that work under low-temperature and evacuated conditions, as shown in Figure 21b. In previous low-temperature devices, sample rotation has only been possible on one axis and, consequently, a few of the Bragg peaks required for structural analysis were immeasurable. Using the new sample-rotation system, almost all peaks required for structural analysis became measurable by SENJU, even at low temperatures [29]. The size of a sample crystal that can be measured using SENJU is compatible with that for measurements of various types of physical properties. This means that the development of SENJU has facilitated the determination of both crystal structure and physical properties with the same sample under the same condition. We believe that SENJU will open up a new frontier of material science using single-crystal neutron diffraction.

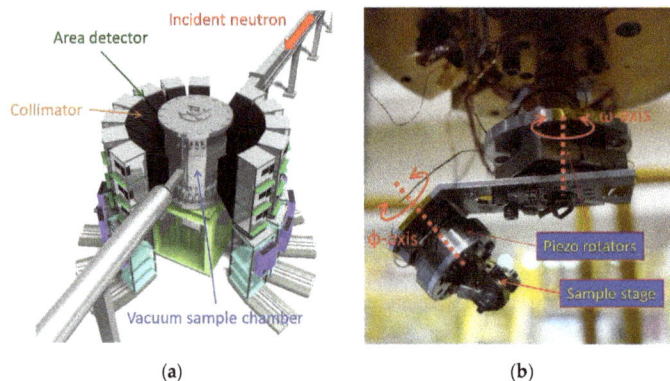

(a) (b)

Figure 21. (**a**) Schematic view of SENJU. The sample crystal is set in a vacuum sample chamber. Scattered neutrons are detected by area detectors arranged around the chamber. (**b**) Sample rotation device for SENJU. A sample crystal fixed on an aluminum rod is set on the sample-mounting table. The sample can be rotated with along the φ and ψ-axes by using piezo-rotators.

Table 14. Specifications of SENJU.

Moderator	Poisoned decoupled
Wavelength of incident neutron	0.4–4.4 Å (1st frame) 4.6–8.8 Å (2nd frame)
Neutron intensity (@1 MW)	1.3×10^6 n/s/mm^2
$L_{moderator\text{-}sample}$	34.8 m
$L_{sample\text{-}detector}$	800 mm
Solid angle of detectors	30.2% for 4π
Detector covered region	13.0–167.0°
Detector size	256 mm $\times$ 256 mm
Detectors pixel size	4.0 mm $\times$ 4.0 mm
Number of detectors	37

4. Nano-Structure Group Instruments

In the nano-structure group, four instruments are involved. These include a small- and wide-angle neutron scattering instrument TAIKAN (BL15), horizontal geometry reflectometer SOFIA (BL16), vertical geometry reflectometer SHARAKU (BL17), and total scattering diffractometer NOVA (BL21). These are not simply built based on existing instruments. TAIKAN has a large area detector bank to cover not only small scattering angle region but also larger angles, which can maximize the advantage of a pulsed neutron source. Moreover, TAIKAN is equipped with advanced neutron optics to fully utilize polarized neutrons for investigating magnetic structure in nano-materials. The ability to perform polarized neutron experiments is one of the key future requirements also at SHARAKU, because investigation of magnetic interface is an important task for this instrument. New approaches are being tried at NOVA and SOFIA. At SOFIA, an approach to measuring geometry by using a combination of a focusing supermirror and PSD is under development to increase the measurement efficiency. At NOVA, as one of attempts to solve the complex and long-standing problem of scan locus, a Fermi chopper has been installed.

4.1. Small- and Wide-angle Neutron Scattering Instrument, TAIKAN

The small-angle neutron scattering (SANS) technique has been indispensable in research on microstructures, higher-order structures, and hierarchical structures in materials science and life science. However, progress in nanotechnology and research on complex multi-component or multi-phase systems and nonequilibrium systems has created the need to endow the SANS instrument with the ability to measure structural information more efficiently with higher structural and temporal resolutions.

In order to meet these needs, the small- and wide-angle neutron scattering instrument, TAIKAN, has been developed and used at beamline BL15 in MLF within J-PARC since JFY2009 [30]. TAIKAN is designed to cover a wide Q-range (Q = 0.0005–20 Å^{-1}) by using neutrons over a broad wavelength range (0.5–8 Å) produced at a coupled supercritical hydrogen moderator at the spallation neutron source in MLF, and its detector geometry covers a wide solid angle with the good connectivity of the scattering angle. TAIKAN is composed of the following set of components: a beam shutter, optical devices (neutron guide tubes, slits, collimators, multi-channel supermirror polarizer, and advanced magnetic beam focusing device), T0 chopper, three disk choppers, sample stage, vacuum scattering chamber, five detector banks, and beam stopper. On the small-angle and the middle-angle detector banks in the vacuum chamber, and the high-angle and the backward detector banks in atmosphere, more than 1500 ^{3}He PSD tubes with 8 mm in diameter and about 0.6 MPa in ^{3}He gas pressure have been installed. On the ultra-small-angle detector bank in the vacuum chamber, a high-resolution scintillation detector with an active size of 127 mm in diameter and spatial resolution of about 0.5 mm has been installed. Figure 22 shows the scattering cross section of mesoporous silica obtained in the Q range of 0.0008–17 Å^{-1} by using TAIKAN. The Q_{min} value is extended down to 0.0005 Å^{-1} after combination with the beam-focusing device, which is a multiple multipole magnet system (quadrupole-spin flipper (SF)-sextupole-SF-sextupole-SF system) and can focus neutrons in the wavelength bandwidth of about 2.5 Å on the ultra-small-angle detector. Characteristic parameters of TAIKAN are listed in Table 15.

TAIKAN has been utilized by many users in various scientific fields, such as soft matter science, chemistry, biology, physics of magnetism, and steel science. To respond to their needs for efficient measurements under various sample conditions, a sample changer, a rheometer, tensile load cell, cryostats, electromagnets, and furnaces are available.

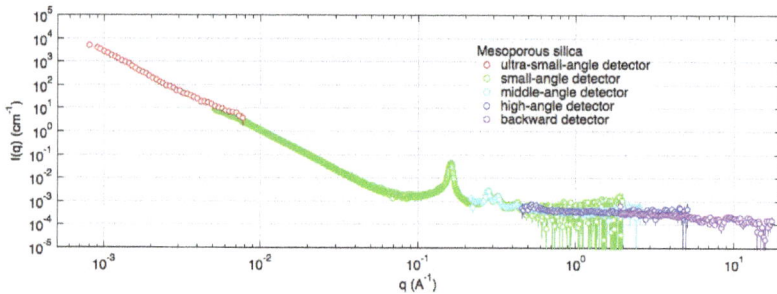

Figure 22. Scattering profile of mesoporous silica measured using TAIKAN.

Table 15. Specifications of TAIKAN.

Beamline	BL15
Moderator	Coupled hydrogen moderator
Wavelength range	0.5–8 Å (unpolarized neutron) 2–8 Å (polarized neutron)
Q range	5×10^{-4}–20 Å^{-1} (unpolarized neutron) 5×10^{-4}–5 Å^{-1} (polarized neutron)
Q resolution	$\Delta Q/Q = 0.3$ @ $Q = 10^{-2}$ Å^{-1}, 4.5×10^{-2} @ $Q = 10^{-1}$ Å^{-1}, 1.6×10^{-2} @ $Q = 1$ Å^{-1}, 3×10^{-3} @ $Q = 10$ Å^{-1}
Detector system	^{3}He 1D-PSD [$\phi = 8$ mm, $L = 300, 500, 600, 800, 1000$ mm (effective length)] 2272 tubes after full installation (1512 tubes are currently installed), ZnS/^{6}LiF scintillation 2D-ultra-small-angle detector [Spatial resolution: 0.5 mm]
Beam monitor	N$_2$ monitor [Efficiency: 10^{-5} @ 1 Å, 64 mm $\times$ 64 mm $\times$ 12 mm (effective volume)]
Beam transport	3Q_c Ni/Ti supermirror guide tube
Polarizer	4.5Q_c Fe/Si supermirror 4-channel V-cavity
Focusing device	Quadrupole magnet [dB/dr = 2.25×10^2 T/m], 1st sextupole magnet [dB/dr^2 = 5.22×10^4 T/m^2], 2nd sextupole magnet [dB/dr^2 = 2.43×10^4 T/m^2], Resonance spin flippers
Sample stage	Diameter: 700 mm, Weight limit: 750 kgf, Beam level above the stage: 350 mm

4.2. Soft Interface Analyzer, SOFIA

Neutron reflectometry is very useful for investigations of the structures of surfaces and buried interfaces composed of soft materials. Beamline 16 (BL16) of MLF is dedicated to a horizontal-type neutron reflectometer, in which free surface such as air–water interface can be measured using two downward neutron beams (2.22° and 5.71°). We started to accept the neutron beam onto the ARISA reflectometer relocated from the KENS neutron facility of KEK in 2008 tentatively, and replaced it with a brand-new reflectometer SOFIA (SOFt-Interface Analyzer) in 2011 through collaboration between JST/ERATO and KEK [31,32]. Owing to the high-flux beam of the J-PARC, the exposure time for one measurement is drastically shortened in comparison with that at various neutron facilities in Japan, such as the JRR-3 research reactor and the KENS spallation source. In addition, the accessible reflectivity is extended to 10^{-7} by careful background treatment, which is approximately two orders of magnitude lower than that of the reflectometers in JRR-3 and KENS.

Table 16 lists typical samples, sample size (and beam size), accessible Q range, and exposure time for full Q-range measurement at the SOFIA reflectometer with proton power of 500 kW. Table 17 lists the specifications of SOFIA. Even though the beam power is half of the planned value, 1 MW, the time required for one measurement is short and competitive with other state-of-the-art reflectometers owing to the high-flux beam of J-PARC. Hence, SOFIA is equipped with a sample changer capable of large quantities of samples, for example, 28 substrates with a diameter of 2-inch, for high throughput measurements. To take further advantage of the high-flux beam, SOFIA can change the repetition rate of the neutron pulses for use of a wide wavelength band from 0.2 nm to 1.76 nm by chopping every two neutron pulses, and it is equipped with sample environments for time-slicing measurement such as liquid injection system, temperature-jump system, and potentiometer [102,103]. These enable us to measure structural evolution over time slices of a few seconds to tens of minutes with a wide Q-range in the swelling process of polymer thin film, thermal diffusion by temperature annealing, electrochemical reactions in the charging/discharging process, and so on. As SOFIA employs an event-recording system similar to that in other instruments in MLF, the time resolution of kinetics measurement can be changed arbitrarily after the measurement considering the balance of data statistics and the speed of the structural change.

With these features, more than 60 reviewed papers and proceedings have been published using BL16 in various research fields such as polymer science, interfacial science, material science, electrochemistry, and tribology.

Table 16. List of typical samples, sample size (and beam size), accessible Q range, and exposure time required for full Q-range measurement at SOFIA reflectometer with proton power of 500 kW.

Sample Interface	Sample Size (Beam Size)	Q Range	Exposure Time
air/Si	50.8 mm Ø (30 mm × 40 mm)	<2 nm^{-1}	$\frac{3}{4}$ h
air/protonated polymer/Si	50.8 mm Ø (30 mm × 40 mm)	<2 nm^{-1}	1 h
air/deuterated polymer/Si	50.8 mm Ø (30 mm × 40 mm)	<4 nm^{-1}	$\frac{1}{2}$ h
air/D_2O	25 mL (40 mm × 40 mm)	<2 nm^{-1}	$\frac{1}{4}$ h
air/no reflection water	25 mL (40 mm × 40 mm)	<1 nm^{-1}	3 h
Si/D_2O	76.2 mm Ø (30 mm × 50 mm)	<2 nm^{-1}	1 h
Si/protonated polymer/D_2O	76.2 mm Ø (30 mm × 50 mm)	<2 nm^{-1}	1 h
Si/deuterated polymer/H_2O	76.2 mm Ø (30 mm × 50 mm)	<2 nm^{-1}	1 h

Table 17. Specifications of SOFIA reflectometer.

Wavelength	0.2–0.88 nm (single frame) 0.2–1.7 nm (double frame)
Incident angle	<6°
Q range	<6 nm^{-1} (depend on reflectivity of a sample)

4.3. Polarized Neutron Reflectometer, SHARAKU

Neutron reflectometry has been used widely to investigate multi-layered structures and interfaces inside artificial thin films. Neutron reflectometry probes a nanometric-layered structure with the contrast of not only the nuclear scattering length density (SLD) but also the magnetic SLD. Therefore, neutron reflectometry allows for the analysis of magnetic structures in magnetic thin films [33]. The polarized neutron reflectometer SHARAKU was constructed at BL17 in MLF. SHARAKU uses polarized neutrons, which can enhance sensitivity to magnetic elements in thin films.

SHARAKU is equipped with a spin polarizer, spin analyzer, and two spin flippers inserted into the BL on demand because the spin polarization states of both the incident and the reflected neutrons must be determined for polarized neutron reflectivity measurement with full polarization analysis. The cold neutrons from the coupled moderator of MLF are passed through the polarizer, which consists of polarizing Fe/Si supermirrors, to polarize the neutron spin. The spin polarity can be switched using a non-adiabatic two-coil spin flipper with a flipping ratio of 40–80. The spin state of the neutrons reflected from the sample is analyzed using a set of stacked Fe/Si supermirrors for full polarization analysis measurements.

In the specular reflectivity profile measurement, a ^{3}He gas tube detector is used. Both high sensitivity to neutrons and insensitivity to background noise of the ^{3}He detector allows for neutron reflectometry measurements down to the reflectivity order of 10^{-6}. SHARAKU enables us to use a 2D PSD and a multi-wire proportional neutron counter (MWPC) with a detection area of 128 mm × 128 mm, the pixel size of 0.5 mm × 0.5 mm, and spatial resolution of 1.8 mm. MWPC is used for off-specular neutron reflectivity and SANS measurements, including grazing-incidence SANS. The off-specular reflectivity measurements reveal not only the layered structure in the depth direction but also the lateral structure in the film plane of a sample. Table 18 summarizes the specifications of SHARAKU.

Table 18. Specification of SHARAKU.

Wavelength	0.24–0.88 nm (polarized neutron)
	0.11–0.88 nm (unpolarized neutron)
Scattering angle	0–18°
Maximum Q	8.19 nm^{-1} (polarized neutron)
	17.9 nm^{-1} (unpolarized neutron)

The sample is mounted vertically on a sample holder set on a rotating table with a diameter of 700 mm and a load capacity of 900 kgf; therefore, it has large flexibility for a variety of sample environment equipment. A 7T superconducting magnet and a 1T electromagnet can be used for the measurement of magnetic films. The magnetic field axis can be switched between the directions parallel and normal to the sample plane. A refrigerator system allows for neutron reflectometry measurement in the temperature range of 4–300 K. Sealed sample cells are equipped for measurement at the liquid-solid interface in a solution and measurements under controlled atmosphere.

The polarized neutron reflectometry experiment is a powerful method for the analysis of magnetic structure of artificial thin films; however, the research fields in the user experiment proposals are not limited to solid-state physics of magnetic materials. SHARAKU has accepted a wide variety of experiments, for instance, polymer thin films and electrochemistry.

4.4. High-Intensity Total Diffractometer, NOVA

Total scattering is a technique to analyze non-crystalline structure in materials by obtaining the real-space atomic-pair correlation function by Fourier transform of diffraction profile. NOVA was designed to perform total scattering, and it is the most intense powder diffractometer with reasonable resolution ($\Delta d/d \sim 0.5\%$) in J-PARC. To realize high statistical accuracy in the high-Q range ($Q = 2\pi/d$, where d is a lattice constant), which has a sever effect on the reliability of the pair-correlation function, the neutron flight pass between the neutron source and the sample was shortened (15 m) and the wavelength bandwidth of incident neutrons was widened (~8 Å). Consequently, NOVA covers a wide Q range, 0.01 Å$^{-1} \leq Q \leq 100$ Å^{-1}, in one measurement. Based on the high neutron flux of J-PARC, real-time observation of non-equilibrium states is feasible. The construction of NOVA was supported by the New Energy and Industrial Technology Development Organization (NEDO) under Advanced Fundamental Research Project on Hydrogen Storage Materials ("HydroStar") and is open to users since 2012. Characteristic parameters of NOVA are listed in Table 19.

One of the main objects of NOVA is to determine the mechanism of hydrogen storage in materials. To observe hydrogen absorption and desorption, NOVA is equipped with in situ sample environments such as H_2/D_2 gas atmosphere (up to 10 MPa, temperature can be controlled from 50 K to 473 K). The sample environments available in NOVA are listed in Table 20. Vanadium foil furnace (RT–1373 K) and impedance measurement apparatus (temperature: $\leq$ 550 K, freq.: 4 Hz–1 MHz) will be ready as soon as a radial collimator is installed. The cryogenic system of the common sample environment equipment in MLF is also expected to be ready as soon as the collimator is ready. Another unique feature of NOVA is its capability of inelastic measurement by using a Fermi chopper. A study is on-going for correction of the self-term cross section of ^{1}H in $S(Q)$ by measuring $S(Q, E)$.

A variety of scientific fields have been covered by NOVA: hydrogen storage materials (for example [104,105]), battery materials containing amorphous and liquid phases (for example [106,107]), and magnetic structure analysis (for example [108]). The capability of longer d-range measurement is useful for magnetic structure analysis.

Table 19. Specifications of NOVA.

Beamline	BL21		
Moderator	Decoupled hydrogen moderator		
	Detector bank	$\Delta Q/Q$ [%]	Q-range [Å^{-1}] (d-range[Å])
	Small-angle	4–50	0.03–8 (0.8–209)
Q range and resolution	20°	1.7–3.9	0.2–26 (0.2–31)
	45°	0.9–1.5	0.4–50 (0.1–16)
	90°	0.5–0.7	1–82 (0.08–6.3)
	High-angle	0.3–0.35	1.4–100 (0.06–4.5)
Detector system	^{3}He 1D-PSD [ϕ = 1/2 inch, L = 800 mm (effective length)]		
Beam monitor	Gas Electron Multiplier (B converter, 0.1% efficiency)		
Beam Size	Typical size: 6 mm width × 20 mm height		
	Beam size can be changed from 5 mm×5 mm to 20 mm×20 mm		

Table 20. List of available sample environments in NOVA.

Apparatus	Purpose	Specification
Sample changer	Automatic sample exchange	Samples per load: 10 or 40
Temperature controlled sample changer	Automatic sample exchange and temp. control	Samples per load: 18 Temperature : 20–700 K
Top load cryostat	Low temperature	Temperature : 5–700 K
Vanadium furnace	High temperature	Temperature : RT–1373 K
Hydrogen pressure-composition-temperature measurement	H_2 ab/desorption	Temperature : 50–473 K Gas pressure : $\leq$10 MPa H_2/D_2

5. Pulsed Neutron Application Group Instruments

The pulsed neutron application group consists of four unique instruments in various objects other than neutron scattering. ANNRI developed a unique γ-ray spectroscopy method combined with a TOF method. The challenge associated with NOP is the determination of neutron lifetime with an accuracy of 1 s, which will be a great step up from the standard model for elementary particle physics. NOBORU provides opportunities for R&D activities and trial users who have new ideas. RADEN disseminates the power of energy-resolved neutron imaging to the world. The outline of each instrument is described in the following section. ANNRI and NOBORU constitute the JAEA beamlines, NOP is the KEK beamline, and RADEN is the public beamline. ANNRI, NOBORU, and NOP have been open to the user program since 2008, whereas RADEN joined the program in 2015.

5.1. Accurate Neutron-Nucleus Reaction measurement Instrument, ANNRI

Accurate Neutron-Nucleus Reaction measurement Instrument (ANNRI) is used for studies of nuclear science, such as nuclear data for nuclear technology and astrophysics, quantitative analyses, and so on [34].

5.1.1. Instrument Description

In ANNRI, a neutron collimator, neutron filters, and chopper systems are installed to deliver an intense neutron beam with good quality to detector systems. Two detector systems are installed. An array of germanium (Ge) detectors is installed at the flight length of 21.5 m. The other one is a NaI spectrometer located at a flight length of 27.9 m. Using the neutron TOF method and the γ-ray detector systems, both the energy of incident neutrons and the energy of prompt γ-rays from a reaction are obtained.

The array of Ge detectors is composed of two cluster-Ge detectors, eight coaxial-Ge detectors, and anti-Compton shields around each Ge detector. Because each cluster-Ge detector has seven Ge crystals, the array has 22 Ge crystals. The peak efficiency of the spectrometer is 3.6 ± 0.11% for 1.33-MeV rays [109]. The NaI(Tl) spectrometer is composed of two anti-Compton NaI(Tl) scintillators with neutron and γ-ray shields. The cylindrical NaI(Tl) scintillators are located at 90° and 125° with respect to the neutron-beamline, respectively.

5.1.2. Applications

ANNRI is used for neutron-capture gamma-ray measurements in nuclear engineering, elemental analysis and nuclear astrophysics.

5.1.3. Specifications

The typical specifications of the instruments in ANNRI are given in Table 21.

Table 21. Specifications of ANNRI.

Moderator	Coupled Supercritical H_2
Incident neutron energy	$E_n > 0.0015$ eV
Spectrometer	Ge Spectrometer (Flight path length: 21.5 m) NaI Spectrometer (Flight path length: 27.9 m)
Neutron intensity (@sample position @1MW)	@21.5m sample position 4.3×10^7 n/cm^2/s 1.5 meV $< E_n <$ 25 meV 9.3×10^5 n/cm^2/s 0.9 eV $< E_n <$ 1.1 eV 1.0×10^6 n/cm^2/s 0.9 keV $< E_n <$ 1.1 keV
Sample size and/or volume	Using the most downstream collimator, neutron beams with diameters of 22, 15, 7 and 6 mm are provided to suit samples of different sizes.

5.1.4. Highlights

- A novel analytical technique that combines prompt gamma-ray analysis with the TOF technique (TOF-PGA) was developed [110].
- In the neutron-capture cross section measurements of minor actinides, analyses of ^{244}Cm, ^{246}Cm, ^{241}Am, and ^{237}Np have been completed. The results for ^{244}Cm and ^{246}Cm show that neutron-capture cross sections are deduced by using ANNRI, where a small amount (less than 1 mg) of a high radioactive sample can be used [111–113].
- In the neutron-capture cross section measurements of stable isotopes, miss-assigned resonances were found for ^{112}Sn, ^{118}Sn, ^{107}Pd, and ^{142}Nd. These results show that even in the case of a stable isotope, there are many miss-assigned resonances in the evaluated values [114–116].

5.2. Neutron Optics and Fundamental Physics, NOP

Neutron Optics and Physics (NOP) is a beamline for studies of fundamental physics namely: elementary particle, nuclear, and quantum physics.

5.2.1. Instrument Description

The beamline is divided at its upstream into three branches to conduct different experiments in parallel [117]. Each branch can be used to perform several types of physical experiments in parallel. The three branches are follows: Polarized beam branch, which produces polarizing neutrons by using magnetic supermirrors in its bender; Unpolarized beam branch, which produces the highest energy and the most intense neutron flux in these branches with a supermirror bender; and Low-divergence beam branch, which produces small divergence but dense neutron beam.

5.2.2. Applications

Physics experiments using slow neutrons are being carried out. Precise measurement of the neutron lifetime is performed at the polarized beam branch. Pulsed ultracold neutrons (UCNs) by a Doppler shifter are available at Unpolarized beam branch. Low-divergence beam branch is used for testing detectors and optical elements.

5.2.3. Specifications

The typical specifications of the instruments used in NOP are given in Table 22.

Table 22. Specifications of NOP [35].

Moderator	Coupled Supercritical H_2		
Branch	Unpolarized	Polarized	Low-Divergence
Cross section (Y mm × X mm)	50 × 40	120 × 60	80 × 40
Beam flux (n/cm^2/s@1 MW)	$(3.8 \pm 0.3) \times 10^8$	$(4.0 \pm 0.3) \times 10^7$	$(5.4 \pm 0.5) \times 10^4$
Beam Divergence (Y mrad×X mrad)	m = 2 equivalent	23 × 9.4	0.23 × 0.23 *1
Luminance (n/cm^2/str/s@1MW)	—	$(1.8 \pm 0.1) \times 10^{11}$	$(1.0 \pm 0.1) \times 10^{12}$
Polarization	—	94–96%	—

*1 For most intense position. Maximum divergence is about 14 mrad × 2.4 mrad.

5.2.4. Highlights

Ultracold neutrons (UCNs), which have energy less than ~250 neV (velocity of 6.8 m/s), are used for various precision measurements in fundamental physics. A Doppler shifter produces pulsed UCNs at the NOP beamline [118]. It reflects very cold neutrons (VCNs) with velocity of ~136 m/s by using a mirror moving at 68 m/s and produces neutrons with velocities less than 20 m/s. The intensity of UCNs is 160 cps at 1 MW.

5.3. NeutrOn Beam-line for Observation and Research Use, NOBORU

This instrument serves a versatile neutron field for characterizing the neutron source as well as for R&D on various devices, irradiation and analysis of materials, and so on.

5.3.1. Instrument Description

NOBORU was constructed at BL10 in MLF at J-PARC in 2007 by the neutron source group [36]. The primary mission of NOBORU is facility diagnostics to study neutronics performance of the neutron source at MLF [119]. In addition, NOBORU provides a suitable neutron environment for testing various detectors and devices, new ideas for experimental techniques, and irradiation of high-energy (~MeV) neutrons. The instrument has a beam cross-section of 100 mm × 100 mm at a sample position of around 14 m from the neutron source. Figure 23 shows the instrument components of NOBORU.

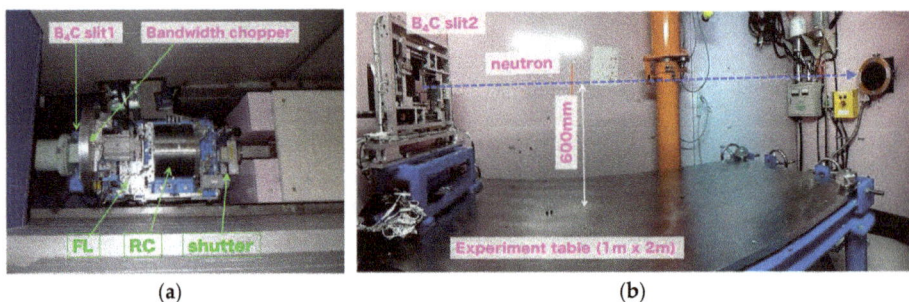

Figure 23. Pictures of NOBORU beamline: (**a**) upstream devices and (**b**) experiment room, where FL and RC refer to filter exchanger and rotary collimator, respectively.

5.3.2. Applications

NOBORU serves as a test beam port to encourage the introduction of innovative instruments into MLF. The key devices installed in instruments of MLF, such as MAGIC chopper [120], wavelength-shifting fiber scintillator detector [121–123], and SEOP system, were developed using NOBORU. The innovative pulsed-neutron imaging instrument, RADEN at BL-22, also uses R&D experiences of the imaging experiment conducted using NOBORU [124–129]. Recent development of white neutron holography enables us to observe the 3D local structure of boron-doped materials.

5.3.3. Specifications

The typical specifications of the instruments used in NOBORU are given in Table 23.

Table 23. Specifications of NOBORU.

Moderator	Decoupled Supercritical H_2
Incident neutron wavelength	$\lambda < 10.5$ Å
Resolution ($\Delta\lambda/\lambda$)	0.35% (minimum)
Neutron intensity (@14.0 m sample position @1MW)	4.8×10^7 n/s/cm^2 (<0.4 eV), 1.2×10^7 n/s/cm^2 (>1 MeV), 1.2×10^6 n/s/cm^2 (>10 MeV)
Beam collimation ratio (L/D)	140, 190, 600 and 1875
Filters	none/Cd 2 mm/acryl 6 mm none/Ta 50 µm, In 50 µm, Cu 2 mm/borosilicate glass 1 mm none/Pb 50 mm/Bi 25 mm none/Bi 50 mm/Pb 25 mm

5.3.4. Highlights

- Studies on the high-field magnetic structure of BiFeO$_3$ [130], multiferroic materials, and magnetoelectric oxides were performed by combining neutron Laue diffraction with pulsed magnetic fields of up to 42 T.
- Energy-resolved neutron imaging was performed on SOFC materials [131], dissimilar metal welds [132], additive manufactured Inconel [133], quenched steel rods [134], bent steel plates [135], and natural gold single crystals [136].
- Neutron detection using a superconducting current-biased kinetic inductance detector (CB-KID) with a superconducting Nb meander line of 1 µm width and 40 nm thickness was demonstrated successfully [137,138].

5.4. Energy Resolved Neutron Imaging System, RADEN

RADEN is the world's first pulsed neutron imaging instrument [37]. By utilizing the nature of short-pulsed neutrons, RADEN enables us to conduct very precise and efficient energy-resolved neutron imaging experiments such as Bragg-edge, resonance absorption, and polarized neutron imaging.

5.4.1. Instrument Description

RADEN (Figure 24) is designed to conduct energy-resolved neutron imaging experiments in addition to conventional neutron radiography and tomography experiments. Fine wavelength resolution and the available wide wavelength/energy range are suitable for Bragg-edge imaging and resonance absorption neutron imaging from the viewpoint of visualizing spatial distributions of crystallographic information such as phase, texture and strain, and elemental and thermal information [139]. The 3D polarization analysis system for the magnetic field imaging is one of the unique features of RADEN [6]. Concerning conventional neutron imaging, a large field-of-view of up to 300 mm × 300 mm and a good spatial resolution below 50 μm offer a suitable environment for non-destructive studies of various objects. Moreover, a wide experimental area and several sample stages equipped with RADEN provide capabilities not only to handle large objects and large sample environments but also to conduct in situ experiments.

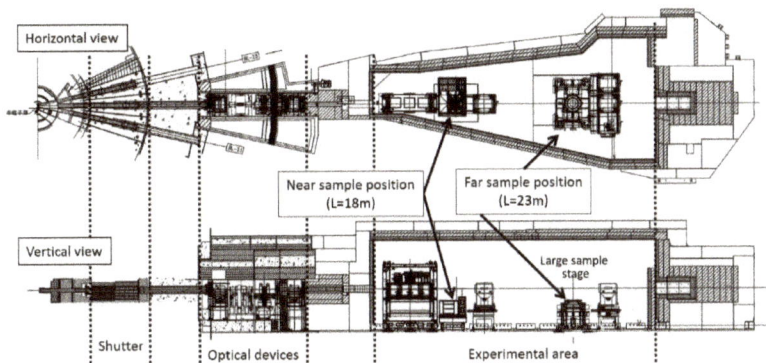

Figure 24. Horizontal and vertical views of RADEN.

5.4.2. Applications

Given that neutron imaging is regarded as a fundamental research technique to visualize internal structures non-destructively, the possible application fields of RADEN are very diverse, including material science, engineering, archeology, biology, agriculture, and industry. From the technical point of view, computational tomography reveals the 3D structural information of an object and provides 2D information of any cross-section, and the movement of inner objects or flow of matter can be also visualized by taking short-instance images using remarkable peak intensity. Furthermore, energy-resolved neutron imaging techniques are used to visualize and quantify 2D distributions of phase, texture, and strain in crystalline samples under process; magnetic fields inside magnetic materials or space; temperature distribution; and segregation of elements.

5.4.3. Specifications

The typical specifications of the instruments used in RADEN are given in Table 24.

Table 24. Specifications of RADEN.

Moderator	Decoupled Supercritical H_2
Incident neutron wavelength	$\lambda < 8.8$ Å ($L = 18$m, 25 Hz) $\lambda < 6.8$ Å ($L = 23$m, 25 Hz)
Resolution ($\Delta\lambda/\lambda$)	0.20% (minimum)
Neutron intensity (@sample position @1MW)	9.8×10^7 n/s/cm^2 ($L/D = 180$) 5.8×10^7 n/s/cm^2 ($L/D = 230$)
Beam size	Maximum 300 mm $\times$ 300 mm @23 m sample position
Detectors	Cooled CCD (2k $\times$ 2k pixels) + ZnS(Li) scintillator, nGEM, μ-NID, Li-glass pixellerated scintillator with multi-anode PMT
Sample environment	Large (load capacity 1 ton), Medium (load capacity 650 kg), Small sample stage (load capacity 10 kg), Polarization analysis system

5.4.4. Highlights

- Visualization of magnetic field of an electric motor in the driving state by polarized pulsed neutron imaging [7].
- Non-uniform charge/discharge property of commercial Li ion batteries.
- Visualization of the phase distribution in a deformed steel sample [135].
- Development of phase imaging technique based on Talbot–Lau interferometer using pulsed neutron beams [140,141].

6. Outcomes from Neutron Instruments at MLF

The world-leading pulsed neutron instruments at MLF in J-PARC are producing cutting-edge scientific data in various research fields, including solid state physics, energy science, materials science including engineering materials, polymer chemistry, soft matter, and geo- and life sciences. The methods using pulsed neutrons have been extended from conventional diffraction of powder samples and single crystals to in situ sample environments and time-dependent measurements (including in-operando experiments); to residual stress mapping measurement and in situ loading technique for engineering materials under heat treatment; from conventional inelastic scattering measurements to overall measurements using multi-E_i techniques covering wide Q-E ranges with multiple energy-resolutions and single-crystal rotation for fully 4D Q-E access; from radiography to energy-resolved neutron imaging, including Bragg-edge mapping analysis, magnetic field imaging, and resonance absorption imaging, reflectometry on free and buried surfaces of matter, and its application to time-resolved reactions at interfaces; and development of the TOF-SANS method and its extension to super-small-angle and wide-angle scattering. In addition to the instrument dedicated to high-pressure science, the introduction of high-pressure accessories to conventional instruments, including diffractometers and spectrometers, is underway. Neutron polarization technique using the SEOP system are ready to be introduced to common experiments in each beamline. Such extensions in experimental techniques are now underway and will help us explore frontier sciences in the near future.

In the following sub-sections, research outcomes in past publications will be highlighted as typical applications developed in the first eight years of MLF. A few of these papers include complementary use results not only between neutron and synchrotron X-ray, but also between neutron and muon beams. The latter is characteristic of MLF.

6.1. Structural Study of Sulfide Super-Ionic Conductor for High-Power All-solid-state Batteries at MLF

Battery technology is a key to energy storage because most mobile and wearable devices are powered by batteries. Lithium-ion battery is still one of the candidates for next-generation batteries. In particular, an all-solid-state battery is expected as the next-generation battery to obtain high power and high energy capacity with safety. For the realization of an all-solid-state battery, a lithium ionic conductor as the solid electrolyte is the key material because poor candidates for solid electrolytes only exist.

The investigations of lithium-ion conductors began in the 1970s. Although many researchers have tried to design and synthesize faster lithium ionic conductors, every lithium conductor has showed lower conductivity than silver and copper ions in solid materials. The efforts to find a good lithium ionic conductor continued until 2011, when a new outstanding lithium ionic conductor, $Li_{10}GeP_2S_{12}$ (LGPS), was discovered by Kamaya et al. [75]. Its ionic conductivity is 1.2×10^{-2} S·cm^{-1}, which is somewhat lower than that of preset organic–liquid electrolytes. This value is incredible, and it propelled us to exchange liquid electrolyte with a solid one. Kanno et al. found many lithium ionic conductors before LGPS. One of them was called Thio-LISICON [142–144]. The materials design concept of crystal solids for lithium-ionic conductors involved the selection of a material system composed of suitable framework structure for ionic conduction. LIthium SuperIonic CONductor, $Li_{14}Zn(GeO_4)_4$ (LISICON) was discovered using this concept [145]. A wide variety of materials with the same type of framework have been synthesized; a wide range of solid solutions formed by aliovalent substitutions introduced to interstitial lithium ions or vacancies have led to high ionic conductivities at elevated temperatures. LISICON is an oxide material, whereas thio-LISICON is a sulfide. The sulfide system has a number of advantages over oxides for constructing ionic conductors according to the above criteria of material design, larger ionic radii and more polarizable character of sulfide ions may improve the mobility of the conducting species.

The composition of LGPS is almost the same as that of the solid solution, $xLi_4GeS_4–(1-x)Li_3PS_4$ as thio-LISICON. By contrast, the crystal structure is quite different, so much so that it can be analyzed as an unknown structure. The crystal structure of LGPS was determined using BL08, SuperHRPD, and synchrotron radiation, as shown in Figure 25. 1D chains are formed by LiS_6 octahedra and $(Ge, P)S_4$ tetrahedra, which are connected by a common edge. These chains are bound by PS_4 tetrahedra. The anisotropic displacement parameters indicate that the vibration of lithium ions is broadened toward the *c*-axis. Figure 26a shows the nuclear scattering length density of lithium ions obtained by the maximum entropy method (MEM). The MEM map also revealed a 1D lithium conduction pathway along the *c*-axis at RT. However, the other MEM map at a higher temperature shows a 3D lithium-ion conduction pathway. It indicates a lithium site that does not contribute to ionic conduction at RT and this site could move in this structural framework at RT if the structure were to be optimized. As a result, the ionic conductivity increases if 3D conduction occurs.

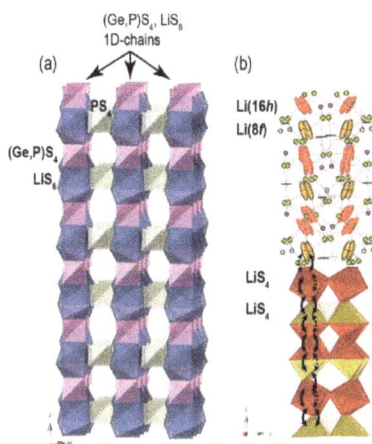

Figure 25. Crystal structure of $Li_{10}GeP_2S_{12}$.

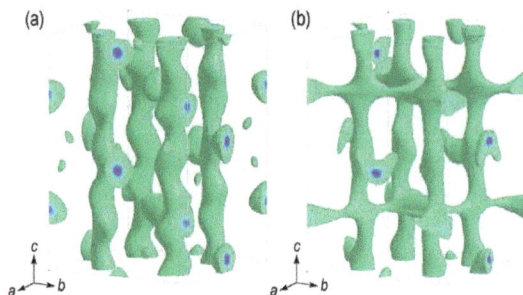

Figure 26. Nuclear scattering length density of lithium atoms in (a) $Li_{10}GeP_2S_{12}$ (LGPS) and (b) $Li_{9.54}Si_{1.74}P_{1.44}S_{11.7}Cl_{0.3}$ (LSPSC).

Kato et al. discovered new lithium-ion conductors with higher conductivity, $Li_{9.54}Si_{1.74}P_{1.44}S_{11.7}Cl_{0.3}$ (LSPSC) and $Li_{9.6}P_3S_{12}$ with the LGPS-type crystal structure [146]. LSPSC is a chlorine-doped material in which silicon is substituted for germanium in the original LGPS system. Its structure was determined by a neutron diffraction experiment using iMATERIA in BL20. The conduction pathway based on MEM analysis is shown in Figure 26b. It indicates a 3D conduction pathway (1D path along c-axis (the same as original LGPS) and a 2D pathway in the ab-plane). The 3D lithium diffusion might be caused by the small amount of chlorine that is substituted mainly for sulfur in PS_4, which connects the 1D chains formed by LiS_6 octahedra and $(Ge,P)S_4$ tetrahedra. As a result, its conductivity increases to double the conductivity of LGPS. The conduction mechanism in LGPS-type structures is also being investigated using a quasi-elastic technique at BL02, DNA. In the near future, details pertaining to the lithium-ion conduction mechanism will be clarified and a new concept of lithium-ionic conductors determined from structural information and dynamics will be designed and synthesized. This cycle of design, synthesis, and structure determination is important to develop new ionic conductors and other functional materials.

These line-ups of instruments in J-PARC MLF are very useful for not only material science but also for the application of these materials. For instance, SPICA, a special environment neutron powder diffractomter dedicated to battery study was built on BL09. SPICA is used to observe battery reactions as structural changes in active materials contained in commercial batteries under operation. SPICA will be used to conduct material science studies in practical devices. All-solid-state batteries using LSPSC as the electrolyte have already been developed [146] and they show the features of both batteries, which serve as electric storage, and capacitors, which have high current density. That kind of reaction mechanism in devices is still in the black box. Neutron scattering will be a key technique for improvement of battery devices among others.

6.2. Neutron Diffraction Study of Phase Transformations in Steels—Effect of Partial Quenching on Phase Transformation in Nano-Bainite Steel

Neutron diffraction is well known as one of the powerful probes in engineering materials studies because of the high penetration ability of neutrons. The applications of neutron diffraction, which were limited in studies related only to residual stresses in engineering components, have been expanded to many types of in situ measurements to evaluate deformation and functional behaviors of various engineering materials with increasing neutron beam intensity and instrument quality. Careful analysis of the Bragg peaks in a neutron diffraction pattern can reveal important structural details of a sample material such as internal stresses, phase conditions, dislocations, and texture.

Nano-bainite steels were designed by alloying to form a bainite structure in the low-temperature region of 473–673 K after austenization. These steels with a microstructure comprising nanoscaled laths/plates of bainitic ferrite and carbon-enriched film austenite exhibit tensile strength greater than 2 GPa and fracture toughness of 30 MPa m$^{1/2}$. However, their extremely slow bainite transformation rate has hindered their use in engineering applications. Therefore, accelerating the bainite transformation is a necessary for further application of these steels. To that end, we focused on a partial quenching and bainite transformation (QB) treatment. An in situ neutron diffraction technique was employed to clarify the effect of partial quenching on the following bainite transformation [98].

The in situ neutron diffraction experiments were performed using the engineering neutron diffractometer, TAKUMI, at J-PARC, along with two heat-treatment processes: with and without partial quenching before isothermal holding at 523 K or 573 K. The chemical composition of the steel used in this study was Fe–0.79C–1.98Mn–1.51Si–0.98Cr–0.24Mo–1.06Al–1.58Co (wt %). The diffraction profiles obtained were analyzed using the Z-Rietveld code to determine the phase fraction and the lattice parameter of each constituent. The convolutional multiple whole profile (CMWP) method was employed for profile analysis to determine dislocation density and substructure.

The heat treatment route and the corresponding dilatometry curves are shown in Figure 27a. Figure 27b shows that when the specimen was cooled from 1173 K to 350 K, that is, partial quenching, the austenite phase (face-centered cubic, fcc) transformed into the martensite phase (body-centered tetragonal, bct). After reheating and holding the specimen at 523 K, the growth of bcc (body-centered cubic) peaks together with the decreased intensity of the austenite peaks implied the presence of bainite transformation. Figure 27c,d shows a comparison of the kinetics of a direct isothermal bainite transformation (DIT) and the QB treatments at 573 K and 523 K, respectively. As can be seen, the bainite transformation occurs more quickly under QB treatment than it does under DIT. A comparison of the austenite diffraction peaks around the martensite transformation (collected at the time marked by blue points in the heat-treatment route in Figure 27a) shows that an apparent peak broadening occurred after the martensite transformation. The CMWP analysis showed that high-density dislocations of 1.51 × 10^{15} m^2 were introduced into austenite in the process of the martensite transformation. The dislocations introduced in austenite by accommodating the shape strain of the martensite transformation are believed to have assisted the following bainite transformation.

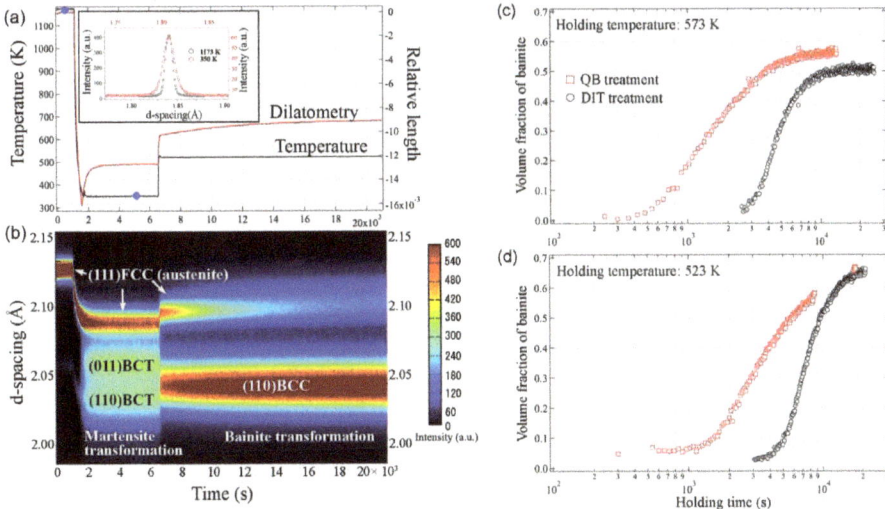

Figure 27. (**a**) Heat treatment route, dilatometry curve, and comparison of (2, 0, 0) peaks of austenite; (**b**) Evolution of diffraction profiles. Comparison of bainite transformation kinetics between the direct isothermal bainite transformation (DIT) and partial quenching and bainite transformation (QB) treatment at isothermal temperatures of (**c**) 573 K or (**d**) 523 K.

The in situ neutron diffraction technique is useful for observing structural changes in steel during heat treatment, which is difficult to achieve using conventional methods. In the future, neutron diffraction techniques are expected to play an important role in studies on steels.

6.3. Neutron Structure Analysis of a [NiFe]hydrogenase-mimicking Complex at BL03 iBIX

Single-crystal neutron diffraction is one of the most fundamental and powerful techniques to determine the arrangement of light elements and magnetic moments of crystalline materials with high accuracy and reliability. Thus, this technique has been used widely and has the potential to be an irreplaceable analytical tool. Recently, two single-crystal neutron diffractometers, iBIX and SENJU, were built at MLF. Specifications and sample environment of iBIX are optimized for protein crystallography and those of SENJU for material physics and chemistry. Consequently, structural studies in various scientific fields are covered by those two diffractometers.

Hydrogen gas is thought of as a promising clean and sustainable energy carrier. Nickel-iron hydrogenases ([NiFe]H$_2$ases), natural enzymes produced by some bacteria and algae, are known to catalyze the transfer of electrons from hydrogen gas (H$_2$) and activate H$_2$ under ambient temperature and pressure conditions. Consequently, [NiFe]H$_2$ases are studied intensively in the field of energy science. In those studies, synthesis of small metal complexes mimicking the catalytic activity of [NiFe]H$_2$ases has been one of the hottest topics for both understanding the catalytic mechanism and the industrial applications of [NiFe]H$_2$ases.

In 2007, Ogo et al. reported a nickel—ruthenium model complex that can mimic the chemical functions of [NiFe]H$_2$ases [147]. In this report, a single-crystal neutron diffraction study of the [NiRu] model complex using the BIX-3 diffractometer at the JRR-3 research reactor showed that this complex has Ni(μ-H)Ru, a three-center Ni-H-Ru bond. This result strongly suggested that natural [NiFe]H$_2$ases also have the Ni(μ-H)Fe structure in their active sites. However, the use of Ru, an expensive and rare element, was a bottleneck for industrial application.

Recently, Ogo et al. successfully synthesized a functional model complex which has the same [NiFe] core (Figure 28) as [NiFe]H$_2$ases. To confirm existence of the Ni(μ-H)Fe structure at the center of this complex, they carried out single-crystal neutron structure analysis using the iBIX diffractometer in MLF, J-PARC. Given that this complex is unstable at ambient temperature, a single crystal of the complex was quickly mounted on a loop-mount pin and transferred into a cold stream of N$_2$ gas in iBIX. This sample-mounting procedure is similar to that of a conventional single-crystal X-ray diffractometer. The size of the sample crystal was 2.0 mm × 1.0 mm × 0.5 mm, and the measurement temperature was 120 K. The m-H atom ($b_c = -3.7409$) between Ni and Fe was replaced with D ($b_c = 6.674$) in the synthesis process to observe nuclear scattering-length density in a Fourier map clearer. Even though the quality of the sample crystal was not up to the mark for single-crystal diffraction measurement and the Bragg reflections were weak and broadened, 2840 unique reflections were obtained after 5 days measurement at iBIX. The data obtained was used for structure refinement and calculations of Fourier maps.

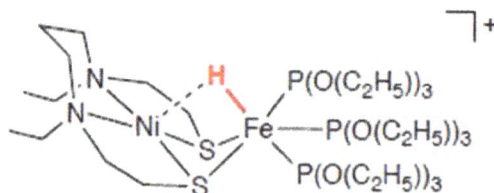

Figure 28. Chemical diagram of [NiFe]H$_2$ase-mimicking functional model complex, which has a [NiFe] core.

Figure 29 shows a difference Fourier map (Fo-Fc map) obtained using a structural model in which the μ-D atom was excluded and a Fourier map (Fo map) calculated using a model in which the μ-D atom was included. In both maps, positive neutron scattering-length density was observed between the Ni and Fe atoms. Although the final *R* value (0.2581 for 931 I > 2σ(I) reflections) was higher than that in typical single-crystal neutron structure analysis of a metal complex because of the quality of the sample crystal, these Fourier maps strongly suggested that the [NiFe] model complex has the expected Ni(μ-D)Fe structure [148].

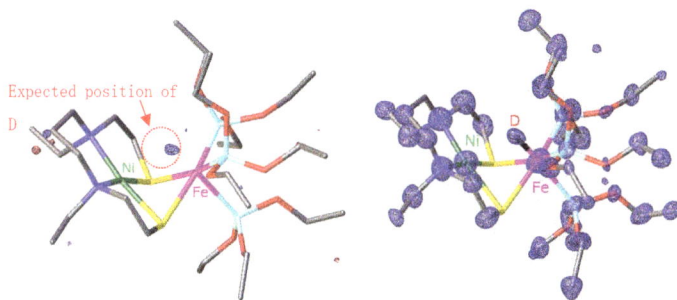

Figure 29. Fourier maps calculated using single-crystal neutron diffraction data. (**left**) Fo-Fc map with a structure model in which the deuterium atom was excluded. Positive scattering-length density is observed at the expected position of the deuterium atom between Ni and Fe. (**right**) Fo-map with a model in which the deuterium atom was included.

iBIX was originally designed for structure analyses of bio-macromolecules, and its sample environment was optimized for fragile and air-sensitive protein crystals. This result showed that iBIX is suitable not only for protein crystals but also for molecular crystals that are unstable at ambient temperature.

6.4. Characterization of Zwitterionic Polyelectrolyte Brush Swelled in Water by SOFIA Reflectometer

It is well known that a surface or interface at which different kinds of materials strongly interact with each other sometimes shows certain unique characteristics, and the behavior is quite different from that of bulk materials. Specifically, interfaces composed of soft materials such as polymers present characteristic features because of the hierarchical structure ranging from nm to μm, in which case, neutron reflectometry is a powerful tool to investigate the depth distribution of polymer species owing to deuteration labeling method.

Polymer brushes, surface-tethered polymers on a solid surface with sufficiently high grafting density, made of polyelectrolyte have attached considerable attention because of their excellent wettability [149], antifouling behavior, and low friction [150] in water. Because these features arise from the highly swollen structure of polyelectrolyte brushes in water, salinity in water is expected to cause changes in the structure because of the interaction between the polyelectrolyte and the ions and strongly affect the surface property. The relationship between the structure and surface property has, however, not been studied well owing to limitations of the method used to characterize nanoscale structures at the interface between the solid surface and the brush and water. This issue is crucial for the application of polyelectrolyte brushes, such as water-based lubrication of a biological surface such as an articular cartilage of mammalian joints. Based on the above background, a neutron reflectivity study using the SOFIA reflectometer [31,32] was performed to investigate the effect of NaCl on the swollen brush structure in water for two twitterionic polyelectrolyte brushes made of phosphorylcholine-containing biomimetic polymer, poly(2-methacryloyloxyethyl phosphorylcholine) (PMPC), and sulfobetaine polymer poly[3-(N-2-methacryloyloxyethyl-N,N-dimethyl)-ammonatopropanesulfonate] (PMAPS) [151].

Figure 30a shows the NR curves of PMPC brush interfaces in contact with D_2O solutions at several NaCl concentrations, as well as the corresponding fits calculated based on the neutron SLD profiles along with distance from the surface. The SLD profile was assumed to be a four-layer model consisting of a quartz substrate, an initiator layer, a gradient brush layer, and D_2O layer. With the SLD profiles of the brush layer, volume fraction of the brush, ϕ, with respect to the distance from the substrate was calculated, as shown in Figure 30b. The obtained volume fraction profile agrees with the self-consistent field theory, in which the ϕ of a polymer brush with a uniform chain length in a good solvent at a position z from the substrate is represented by a parabolic function $\phi = \phi_0[1 - (z/h)^2]$, where ϕ_0 is the volume fraction at the substrate and h is the cut-off thickness of the brush. This indicates that the PMPC polymer brush was extended significantly in the direction normal to the substrate; remarkably, this works even in the case of high salt concentrations, that is to say, NaCl does not have any significant effect on the structure. The origin of this characteristic behavior of the PMPC brush in aqueous NaCl solutions is unclear, but this might be caused by unique interactions among phosphorylcholine units and neutral conditions or hydration structure around the brush.

Figure 31a shows the NR curves of the PMAPS brush interfaces in contact with D_2O solutions at several NaCl concentrations and the corresponding fits. In contrast to the PMPC brush, clear fringes were observed in the NR curve of the PMAPS brush in pure D_2O, indicating that the brush formed a sharp interface. As shown in Figure 31b, the volume fraction profile of PMAPS formed a densely concentrated layer near the substrate surface and a relatively swollen layer with a sharp interface. With the increase in NaCl concentration to 0.1 M, the PMAPS brush gradually extended from 70 nm to 80 nm, and the thickness reached 140 nm with further increase in the concentration to 1.0 M. This indicates that the PMAPS brush shrinks in ion-free water because of the strong attractive dipole–dipole interactions between twitterionic groups, and is gradually extended by adding salt ions as a result of a reduction in interactions among betaines.

The NR results revealed the structure of twitterionic polyelectrolyte brushes swollen in water and the effect of external ions on the structure by changing the electrostatic interaction between the brushes. Note here that the degree of water swelling is consistent with the other properties of the brushes: the antifouling behavior on the brush surface and the low friction coefficient of the PMAPS brush were better at higher NaCl concentrations, whereas those of the PMPC brush were independent of the salt concentration and better than those of PMAPS. Of course, having suggested the consistency between the structure and property, further investigations are required to uncover the full scope of the mechanism of their excellent features. This work nevertheless pointed out the crucial influence of the structure of polyelectrolyte brushes on their excellent properties and the potential of neutron reflectometory for nanotechnology-based material development.

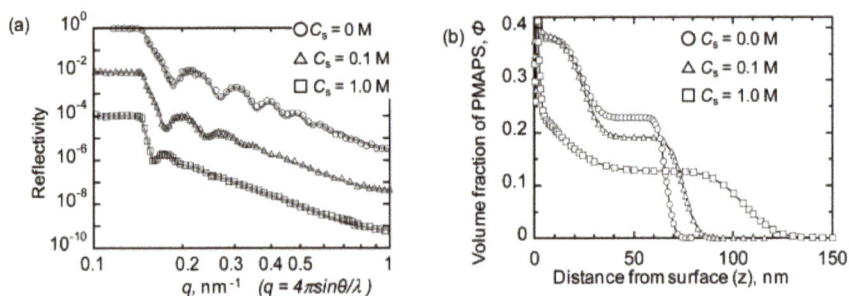

Figure 30. (a) Neutron reflectivity profiles at interface of PMAPS brush–D_2O for different NaCl concentrations and the corresponding fitting curves. (b) Volume fraction of brush with respect to the distance from substrate surface evaluated by fitting (retrieved from [151] with permission from the Royal Society of Chemistry).

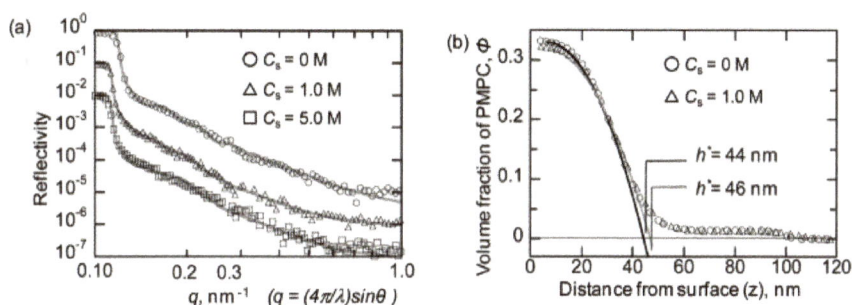

Figure 31. (a) Neutron reflectivity profiles at interface of PMPC brush–D_2O for different NaCl concentrations and corresponding fitting curves. (b) Volume fraction of brush with respect to distance from substrate surface evaluated by fitting (retrieved from [151] with permission from the Royal Society of Chemistry).

6.5. Study of Magnetism in Condensed Matters by Neutron Scattering at MLF—Magnetic Excitations in Hole-Overdoped Iron-Based Superconductors

One of the characteristics of materials science in Japan is that the study of magnetism accounts for a large percentage of the overall number of studies compared to other countries. Needless to say, neutron scattering is a powerful tool to study magnetism. Actually, the Magnetism and Strongly Correlated Electron Systems category accounts for the largest portion of experimental proposals submitted to MLF. As a result, many important and unique outcomes have been produced. For example, magnetic excitations in electron-doped antiferromagnets $Pr_{1.40-x}La_{0.60}Ce_xCuO_4$ were observed up to the sub-eV energy region by using the 4SEASONS spectrometer, which enabled a complementary study with a resonant inelastic X-ray scattering and led to a comprehensive understanding of the doping dependence of magnetic excitations in electron-doped cupper-oxide superconductors [45]. A study of magnetic excitations in the $S = 1/2$ 1D quantum magnet $KCuGaF_6$ over a wide momentum-energy range using the AMATERAS spectrometer unambiguously determined the magnetic exchange interaction and the Dzyaloshinsky–Moriya interaction, the latter of which is the key to unique low-energy elementary excitations under a magnetic field represented by the quantum sine-Gordon model [152]. A detailed study of the ferromagnetic spin waves in the metallic magnet $SrRuO_3$ by means of neutron Brillouin scattering using the HRC spectrometer revealed a unique temperature dependence of the spin gap, which can be attributed to the Berry curvature of the Weyl fermion [57]. A study of the magnetic structure of the honeycomb antiferromagnet $Co_4Nb_2O_9$ using the single-crystal diffractometer SENJU revealed that the magnetic structure is different from that proposed previously, but it better explains the magnetic properties as well as the magnetoelectric response of this compound [153]. In the following, we show in greater detail a recent inelastic scattering study on magnetic fluctuations in iron-based superconductors performed using the 4SEASONS spectrometer [48].

The appearance of the superconducting phase near the antiferromagnetic ordered phase in iron-based superconductors suggest that magnetism plays an essential role in the mechanism of superconductivity. Thus far, a number of inelastic neutron scattering studies on spin fluctuations have been performed to clarify the relationship between magnetism and superconductivity. Previous inelastic neutron scattering studies on electron-doped $Ba(Fe,Ni)_2As_2$ demonstrated that low-energy spin fluctuations are correlated with superconductivity, and they disappear as T_c decreases in the overdoped region [154,155]. By contrast, for hole-doped $(Ba,K)Fe_2As_2$, it was found that low-energy spin fluctuations remain even in heavily overdoped KFe_2As_2, although the superconducting transition temperature T_c is suppressed [156]. To understand the reason for T_c suppression in the hole-overdoped compounds, spin fluctuation studies over the entire Brillouin zone are required. Another important aspect in studies of the magnetism of iron-based superconductors is that it is unclear whether a localized spin picture, which is often successful in describing inelastic neutron scattering results, is always valid as a model for this itinerant system. In fact, many calculations have reproduced the magnetic excitations based on itinerant models [157–159]. Therefore, further experimental examinations of spin fluctuations are required to establish a definite model to describe the magnetism of iron-based superconductors.

Recently, Horigane et al. performed an inelastic neutron scattering study on the spin fluctuations in hole-overdoped $Ba_{1-x}K_xFe_2As_2$ with $x = 0.5$ and $x = 1.0$ by using the 4SEASONS spectrometer [48]. They measured the inelastic scattering spectra of single crystals of $x = 0.5$ (superconducting transition temperature $T_c = 36$ K) and $x = 1.0$ ($T_c = 3.4$ K) at 6 K. Taking advantage of a large quantity of crystals (~5 g) and the high neutron flux at J-PARC, they successfully observed the magnetic excitations in these crystals over entire Brillouin zones.

Figure 32a shows the magnetic excitation spectrum for $x = 0.5$. Spin-wave-like excitations emerging from $K =$ even and extending up to ~200 meV can be observed clearly. The energy scale of the spin-wave-like excitations is similar to that of the excitations observed in the mother compound $x = 0$ and the optimum-doped compound $x = 0.33$ [155]. By contrast, for $x = 1.0$, the bandwidth of the spin-wave-like excitations decreases to ~80 meV (Figure 32b). In a higher-energy region, however, chimney-like excitations with vertical dispersion relations were observed. Less distinct but finite chimney-like excitations exist at $x = 0.5$, indicating that this type of excitation develops by hole doping. Figure 33 shows the hole-concentration (x) dependence of the bandwidth of the spin-wave-like excitations, together with the phase diagram of $Ba_{1-x}K_xFe_2As_2$. The bandwidth is almost constant for $0 < x < 0.5$. However, it shows a sudden decrease at $x > 0.5$, followed by a decrease in T_c. This finding suggests that superconductivity in the hole-overdoped (Ba,K)Fe$_2$As$_2$ is correlated with the effective magnetic exchange interaction J describing the spin-wave-like excitations. By contrast, the chimney-like excitations observed in the high-energy region are hardly described by spin waves based on the localized spin picture. Rather, they resemble the magnetic excitations in antiferromagnetic metals such as Cr [160], $Cr_{0.95}V_{0.05}$ [161], and $Mn_{2.8}Fe_{0.2}Si$ [162]. Therefore, the chimney-like excitations can be attributed to the itinerant nature of the spin fluctuations.

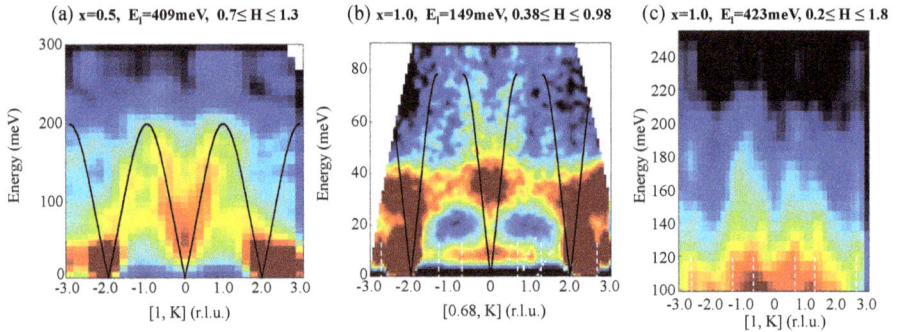

Figure 32. Excitation spectra of $Ba_{0.5}K_{0.5}Fe_2As_2$ and KFe_2As_2 at 6 K observed using 4SEASONS [48]. (a) The spectrum of $Ba_{0.5}K_{0.5}Fe_2As_2$ was measured with incident energy $E_i = 409$ meV; (b,c) Low- and high-energy excitations of KFe_2As_2 measured with $E_i = 149$ meV and $E_i = 423$ meV, respectively. The horizontal and vertical axes show momentum transfers along the K direction in the orthorhombic cell (in reciprocal lattice units) and energy transfers, respectively. The solid lines in (**a,b**) are visual guides, and the dashed white lines in (**b,c**) show the boundaries of the magnetic zone.

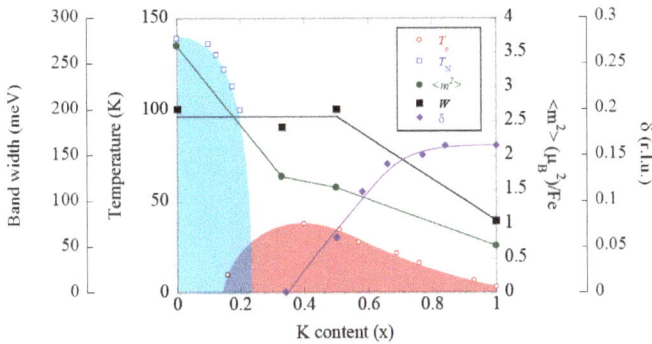

Figure 33. Phase diagram of $Ba_{1-x}K_xFe_2As_2$ [48]. Open circles, open squares, closed circles, closed squares, and closed diamonds show the T_c, Néel temperature T_N, total fluctuating magnetic moment $<m^2>$, bandwidth of the spin-wave-like excitations W, and the incommensurability δ of the magnetic excitations in the reciprocal lattice, respectively.

6.6. Multiple Quantum Beam Investigation of Iron Oxypnictide Superconductor LaFeAsO$_{1-x}$H$_x$

Given that we have not only a neutron source but also a muon source in the same facility, many investigations using both neutron and muon beamlines are being carrying out. Here, we show an example of the outcome of complementary use of neutron and muon beams at MLF.

Quantum beam science involves various types of probe particles available at large facilities. Each probe has its own advantages and disadvantages; a combination of more than one probe may help understand the features of the materials under investigation. One such example is the cooperative use of muon, neutron and synchrotron X-ray radiations with iron oxypnictide superconductor LaFeAsO$_{1-x}$H$_x$ (LFAOHx) [108]. In research of static magnetism, neutrons help with magnetic structure determination, X-rays with the determination of crystallographic structure and symmetry change, and muons help conduct magnetic measurement of powder samples at a considerably faster pace than that possible with other diffraction techniques and help determine magnetic phase diagram as a function of doping concentration x. The muon part of the research is explained in detail in the other part of this article series; here, we focus on the contribution of neutrons in the research on LFAOHx.

Magnetic neutron diffraction from ordered dipole moments shows a simple magnetic form factor $|m_m \times Q|^2$, where the magnetic diffraction vanishes if the scattering vector Q is parallel to the aligned dipole moment m_m and if the magnetic structure is colinear [163]. For LFAOHx, neutron powder diffraction peaks at 10 K and 120 K are compared in Figure 34a, where the appearance of magnetic diffraction peaks is obvious at 10 K. They are indexed as $(1/2, 1/2, L)$ in tetragonal notation after indexing the crystallographic (or nuclear) diffraction peaks. The periodicity of the ordered moments is clear from the position of the magnetic diffractions (Q), but the direction of the ordered moments (m_m) requires additional analysis. We have assumed a few types of magnetic structures and calculated the corresponding diffraction intensities: the best model to account for the observation is the spins aligned in the basal plane at an angle of 45° from the tetragonal a and b axes. The calculated magnetic diffraction profile is shown in Figure 35b. The lower diffraction intensity of the $(1/2, 1/2, 0)$ peak than that of the $(1/2, 1/2, 1)$ is well reproduced. The ordered moment size $|m_m|$ is available as the overall scaling factor between the theoretical profile and the experimental observation.

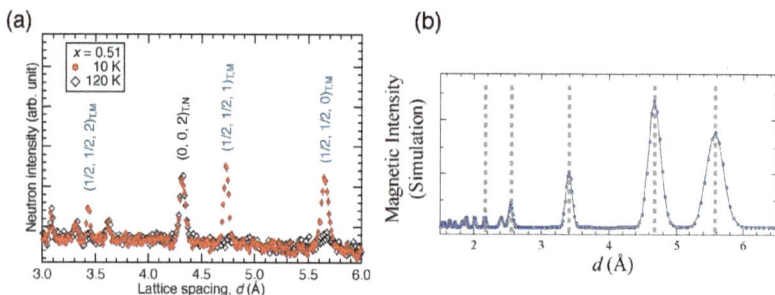

Figure 34. (a) Neutron diffraction profile obtained at SuperHRPD (BL08), MLF, J-PARC (retrieved from [108]); (b) Simulated diffraction intensity of magnetic structure shown in inset of Figure 35.

In LFAOHx, the appearance of the magnetism is related closely to the structural transition clarified by synchrotron X-ray diffraction measurement at KEK-PF. From the asymmetric intensity profile of the split peaks below the structural phase transition temperature (T_S), it is evident that the crystallographic symmetry is reduced to *Aem2* from the higher symmetry of *Cmme* above T_S [108]. With a combination of muon, neutron and X-ray beams, we could comprehensively view the magnetism of LFAOHx and its relationship to superconductivity. This is summarized in a phase diagram shown in Figure 35. In the inset, the magnetic and crystallographic structure for low temperatures is shown ($T < T_N$ and T_S), as determined using the neutron and X-ray beams available at large accelerator facilities.

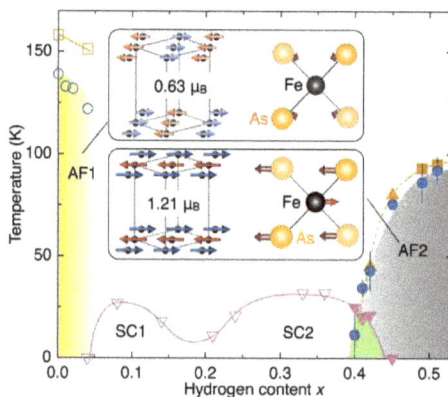

Figure 35. Magnetic and structural phase diagram of LaFeAsO$_{1-x}$H$_x$. The T_N's are determined by muon and partly by neutron beams, and the T_S's by X-ray diffraction. The magnetic and crystallographic structure shown in the inset is determined using neutron beam and X-ray diffraction [108].

Acknowledgments: The authors would like to thank all members of staff who engaged in the development, construction, maintenance and operation of neutron instruments at MLF. The authors acknowledge Masatoshi Arai for his contributions in launching most of the neutron instruments used since the beginning of the J-PARC project, until his retirement from the role of MLF division head in 2015. The authors also acknowledged the former and present MLF division heads, Yujiro Ikeda, Masatoshi Futakawa, and Toshiji Kanaya, for their contributions to MLF management. The neutron scattering experiments using 4SEASONS were performed under the following user programs: Nos. 2012P0201, 2014I0001, 2012B0075 and 2013B0061. The neutron scattering experiments using HRC were approved by the Neutron Scattering Program Advisory Committee of the Institute of Materials Structure Science, High Energy Accelerator Research Organization (Nos. 2009S01, 2010S01, 2011S01, 2012S01, 2013S01, 2014S01, 2015S01 and 2016S01). The POLANO project was approved by the Neutron Scattering Program Advisory Committee of IMSS, KEK (Nos. 2009S09 and 2014S09). The experiments using SOFIA were performed as part of the S-type research project of KEK (2009S08). The neutron scattering experiments using SPICA were approved by the Neutron Scattering Program Advisory Committee of IMSS, KEK (Proposal No. 2014S10, 2009S10). The neutron scattering experiments using SuperHRPD were approved by the Neutron Scattering Program Advisory Committee of IMSS, KEK (Proposal No. 2014S05, 2014S10, 2009S05, 2009S10). The neutron scattering experiments using TAKUMI were performed under user programs Nos. 2014P0100 and 2014I0019. The neutron scattering experiments using NOVA were approved by the Neutron Scattering Program Advisory Committee of the Institute of Materials Structure Science, High Energy Accelerator Research Organization (Nos. 2009S06 and 2014S06). SPICA was supported by the Research and Development Initiative for Scientific Innovation of New Generation Batteries (RISING) project of the New Energy and Industrial Technology Development Organization (NEDO). The construction of the SOFIA reflectometers was a joint project between JST/ERATO and the High Energy Accelerator Research Organization. The construction of NOVA was partially supported by the New Energy and Industrial Technology Development Organization (NEDO) under "Advanced Fundamental Research Project on Hydrogen Storage Materials (HydroStar)". Neutron imaging experiments were partly supported by the Photon and Quantum Basic Research Coordinated Development Program from the Ministry of Education, Culture, Sports, Science and Technology, Japan. RK was supported by JSPS KAKENHI Grant Number JP15K04742. MN was supported by JSPS KAKENHI Grant Number JP26390117. KN was supported by JSPS KAKENHI Grant Number JP24510133. TY and MF were partly supported by JSPS KAKENHI Grant Number JP16H02125. TY was also supported by JSPS KAKENHI Grant Number JP26400376.

Conflicts of Interest: The authors declare no conflict of interest.

References

1. Otomo, T.; Nakajima, K.; Arai, M. (Eds.) *Report of Neutron Instruments Working Group for the Joint Project of KEK and JAERI*; KEK Report 2001-22: Tsukuba, Japan, 2002.

2. Nakajima, K.; Nakatani, T.; Torii, S.; Higemoto, W.; Otomo, T. (Eds.) *Study on a Conceptual Design of a Data Acquisition and Instrument Control System for Experimental Suites at MLF of J-PARC*; JAEA-Technology 2005-005: Tokai, Japan, 2005.

3. Nakatani, T.; Inamura, Y.; Moriyama, K.; Ito, T.; Muto, S.; Otomo, T. Event recording data acquisition system and experiment data management system for neutron experiments at MLF, J-PARC. *JPS Conf. Proc.* **2014**, *1*, 014010. [CrossRef]

4. Nakamura, M.; Kajimoto, R.; Inamura, Y.; Mizuno, F.; Fujita, M.; Yokoo, T.; Arai, M. First demonstration of novel method for inelastic neutron scattering measurement utilizing multiple incident energies. *J. Phys. Soc. Jpn.* **2009**, *78*, 093002. [CrossRef]

5. Kawasaki, T.; Ito, T.; Inamura, Y.; Nakatani, T.; Harjo, S.; Gong, W.; Iwahashi, T.; Aizawa, K. Neutron Diffraction Study of Piezoelectric Material under Cyclic Electric Field using Event Recording Technique. In *Proceedings of the 21st Meeting of the International Collaboration on Advanced Neutron Sources (ICANS-XXI), Ibaraki, Japan, 29 September–3 October 2014*; Oku, T., Nakamura, M., Sakai, K., Teshigawara, M., Tatsumoto, H., Yonemura, M., Suzuki, J., Arai, M., Eds.; Japan Atomic Energy Agency: Ibaraki, Japan, 2016.

6. Shinohara, T.; Hiroi, K.; Su, Y.; Kai, T.; Nakatani, T.; Oikawa, K.; Segawa, M.; Hayashida, H.; Parker, J.D.; Matsumoto, Y.; et al. Polarization analysis for magnetic field imaging at RADEN in J-PARC/MLF. *J. Phys. Conf. Ser.* **2017**, *862*, 012025. [CrossRef]

7. Hiroi, K.; Shinohara, T.; Hayashida, H.; Parker, J.D.; Oikawa, K.; Harada, M.; Su, Y.; Kai, T. Magnetic field imaging of a model electric motor using polarized pulsed neutrons at J-PARC/MLF. *J. Phys. Conf. Ser.* **2017**, *862*, 012008. [CrossRef]

8. Russina, M.; Mezei, F. First implementation of Repetition Rate Multiplication in neutron spectroscopy. *Nucl. Instrum. Methods Phys. Res. Sect. A* **2009**, *604*, 624–631. [CrossRef]

9. Nakajima, K.; Ohira-Kawamura, S.; Kikuchi, T.; Nakamura, M.; Kajimoto, R.; Inamura, Y.; Takahashi, N.; Aizawa, K.; Suzuya, K.; Shibata, K.; et al. AMATERAS: A Cold-Neutron Disk Chopper Spectrometer. *J. Phys. Soc. Jpn.* **2011**, *80*, SB028. [CrossRef]

10. Ehlers, G.; Podlesnyak, A.A.; Niedziela, J.L.; Iverson, E.B.; Sokol, P.E. The new cold neutron chopper spectrometer at the Spallation Neutron Source: Design and performance. *Rev. Sci. Instrum.* **2011**, *82*, 085108. [CrossRef] [PubMed]

11. Bewley, R.I.; Taylor, J.W.; Bennington, S.M. LET, a cold neutron multi-disk chopper spectrometer at ISIS. *Nucl. Instrum. Methods Phys. Res. Sect. A* **2011**, *637*, 128–134. [CrossRef]

12. Shibata, K.; Takahashi, N.; Kawakita, Y.; Matsuura, M.; Yamada, T.; Tominaga, T.; Kambara, W.; Kobayashi, M.; Inamura, Y.; Nakatani, T.; et al. The performance of TOF near backscattering Spectrometer DNA in MLF, J-PARC. *JPS Conf. Proc.* **2015**, *8*, 036022. [CrossRef]

13. Itoh, S.; Endoh, Y.; Yokoo, T.; Kawana, D.; Kaneko, Y.; Tokura, Y.; Fujita, M. Neutron Brillouin scattering with pulsed spallation neutron source – spin-wave excitations from ferromagnetic powder samples. *J. Phys. Soc. Jpn.* **2013**, *82*, 043001. [CrossRef]

14. Kajimoto, R.; Nakamura, M.; Inamura, Y.; Mizuno, F.; Nakajima, K.; Ohira-Kawamura, S.; Yokoo, T.; Nakatani, T.; Maruyama, R.; Soyama, K.; et al. The Fermi chopper spectrometer 4SEASONS at J-PARC. *J. Phys. Soc. Jpn.* **2011**, *80*, SB025. [CrossRef]

15. Itoh, S.; Yokoo, T.; Satoh, S.; Yano, S.; Kawana, D.; Suzuki, J.; Sato, T.J. High Resolution Chopper Spectrometer (HRC) at J-PARC. *Nucl. Instrum. Methods Phys. Res. Sect. A* **2011**, *631*, 90–97. [CrossRef]

16. Itoh, S.; Yokoo, T.; Kawana, D.; Yoshizawa, H.; Masuda, T.; Soda, M.; Sato, T.J.; Satoh, S.; Sakaguchi, M.; Muto, S. Progress in High Resolution Chopper Spectrometer, HRC. *J. Phys. Soc. Jpn.* **2013**, *82*, SA033. [CrossRef]

17. Yokoo, T.; Ohoyama, K.; Itoh, S.; Suzuki, J.; Iwasa, K.; Sato, T.J.; Kira, H.; Sakaguchi, Y.; Ino, T.; Oku, T.; et al. Newly proposed inelastic neutron spectrometer POLANO. *J. Phys. Soc. Jpn.* **2013**, *82*, SA035. [CrossRef]

18. Yokoo, T.; Ohoyama, K.; Itoh, S.; Suzuki, J.; Nanbu, M.; Kaneko, N.; Iwasa, K.; Sato, T.J.; Kimura, H.; Ohkawara, M. Construction of polarized inelastic neutron spectrometer in J-PARC. *J. Phys. Conf. Ser.* **2014**, *502*, 012046. [CrossRef]

19. Yokoo, T.; Ohoyama, K.; Itoh, S.; Iwasa, K.; Kaneko, N.; Suzuki, J.; Ohkawara, M.; Aizawa, K.; Tasaki, S.; Ino, T.; et al. Polarized neutron spectrometer for inelastic experiments at J-PARC: Status of POLANO project. *EPJ Web Conf.* **2015**, *83*, 03018. [CrossRef]

20. Hino, M.; Oda, T.; Yamada, N.L.; Endo, H.; Seto, H.; Kitaguch, M.; Harada, M.; Kawabata, Y. Supermirror neutron guide system for neutron resonance spin echo spectrometers at a pulsed neutron source. *J. Nucl. Sci. Technol.* **2017**, *54*, 1223–1232. [CrossRef]

21. Torii, S.; Yonemura, M.; Surya Panca Putra, T.Y.; Zhang, J.; Miao, P.; Muroya, T.; Tomiyasu, R.; Morishima, T.; Satoh, S.; Sagehashi, H.; et al. Super High Resolution Powder Diffractometer at J-PARC. *J. Phys. Soc. Jpn.* **2011**, *SB020*, 1–4. [CrossRef]

22. Yonemura, M.; Mori, K.; Kamiyama, T.; Fukunaga, T.; Torii, S.; Nagao, M.; Ishikawa, Y.; Onodera, Y.; Adipranoto, D.S.; Arai, H.; et al. Development of SPICA, New Dedicated Neutron Powder Diffractometer for Battery Studies. *J. Phys. Conf. Ser.* **2014**, *502*, 012053. [CrossRef]

23. Ishigaki, T.; Hoshikawa, A.; Yonemura, M.; Morishima, T.; Kamiyama, T.; Oishi, R.; Aizawa, K.; Sakuma, T.; Tomota, Y.; Arai, M.; et al. IBARAKI materials design diffractometer (iMATERIA) -Versatile neutron diffractometer at J-PARC. *Nucl. Instrum. Methods Phys. Res. Sect. A* **2009**, *600*, 189–191. [CrossRef]

24. Hattori, T.; Sano-Furukawa, A.; Arima, H.; Komatsu, K.; Yamada, A.; Inamura, Y.; Nakatani, T.; Seto, Y.; Nagai, T.; Utsumi, W.; et al. Design and performance of high-pressure PLANET beamline at pulsed neutron source at J-PARC. *Nucl. Instrum. Methods Phys. Res. Sect. A* **2015**, *780*, 55–67. [CrossRef]

25. Harjo, S.; Ito, T.; Aizawa, K.; Arima, H.; Abe, J.; Moriai, A.; Iwahashi, T.; Kamiyama, T. Current status of engineering materials diffractometer at J-PARC. *Mater. Sci. Forum* **2011**, *681*, 443–448. [CrossRef]

26. Ito, T.; Nakatani, T.; Harjo, S.; Arima, H.; Abe, J.; Aizawa, K.; Moriai, A. Application software development for the engineering materials diffractometer, TAKUMI. *Mater. Sci. Forum* **2010**, *652*, 238–242. [CrossRef]

27. Tanaka, I.; Kusaka, K.; Hosoya, T.; Niimura, N.; Ohhara, T.; Kurihara, K.; Yamada, T.; Ohnishi, Y.; Tomoyori, K.; Yokoyama, T. Neutron structure analysis using the IBARAKI biological crystal diffractometer iBIX at J-PARC. *Acta Cryst.* **2010**, *D66*, 1194–1197. [CrossRef] [PubMed]

28. Kusaka, K.; Hosoya, T.; Yamada, T.; Tomoyori, K.; Ohhara, T.; Katagiri, M.; Kurihara, K.; Tanaka, I.; Niimura, N. Evaluation of performance for IBARAKI biological crystal diffractometer iBIX with new detectors. *J. Synchrotron Rad.* **2013**, *20*, 994–998. [CrossRef] [PubMed]

29. Ohhara, T.; Kiyanagi, R.; Oikawa, K.; Kaneko, K.; Kawasaki, T.; Tamura, I.; Nakao, A.; Hanashima, T.; Munakata, K.; Moyoshi, T.; et al. SENJU: A new time-of-flight single-crystal neutron diffractometer at J-PARC. *J. Appl. Cryst.* **2016**, *49*, 120–127. [CrossRef] [PubMed]

30. Takata, S.; Suzuki, J.; Shinohara, T.; Oku, T.; Tominaga, T.; Ohishi, K.; Iwase, H.; Nakatani, T.; Inamura, Y.; Ito, T.; et al. The Design and *Q* resolution of the Small and Wide Angle Neutron Scattering Instrument (TAIKAN) in J-PARC. *JPS Conf. Proc.* **2015**, *8*, 036020. [CrossRef]

31. Yamada, N.L.; Torikai, N.; Mitamura, K.; Sagehashi, H.; Sato, S.; Seto, H.; Sugita, T.; Goko, S.; Furusaka, M.; Oda, T.; et al. Design and Performance of Horizontal Type Neutron Reflectometer SOFIA at J-PARC/MLF. *Euro. Phys. J. Plus* **2011**, *126*, 108. [CrossRef]

32. Mitamura, K.; Yamada, N.L.; Sagehashi, H.; Torikai, N.; Arita, H.; Terada, M.; Kobayashi, M.; Sato, S.; Seto, H.; Gokou, S.; et al. Novel Neutron Reflectometer SOFIA at J-PARC/MLF for In-Situ Soft-Interface Characterization. *Polymer J.* **2013**, *45*, 100. [CrossRef]

33. Takeda, M.; Yamazaki, D.; Soyama, K.; Maruyama, R.; Hayashida, H.; Asaoka, H.; Yamazaki, T.; Kubota, M.; Aizawa, K.; Arai, M.; et al. Current Status of a New Polarized Neutron Reflectometer at the Intense Pulsed Neutron Source of the Materials and Life Science Experimental Facility (MLF) of J-PARC. *Chin. J. Phys.* **2012**, *50*, 161–170.

34. Kimura, A.; Harada, H.; Nakamura, S.; Iwamoto, O.; Toh, Y.; Koizumi, M.; Kitatani, F.; Furutaka, K.; Igashira, M.; Katabuchi, T.; et al. Current activities and future plans for nuclear data measurements at J-PARC. *Eur. Phys. J. A* **2015**, *51*, 180. [CrossRef]

35. Mishima, K.; Ino, T.; Sakai, K.; Shinohara, T.; Hirota, K.; Ikeda, K.; Sato, H.; Otake, Y.; Ohmori, H.; Muto, S.; et al. Design of neutron beamline for fundamental physics at J-PARC BL05. *Nucl. Instrum. Methods Phys. Res. Sect. A* **2009**, *600*, 342. [CrossRef]

36. Oikawa, K.; Maekawa, F.; Harada, M.; Kai, T.; Meigo, S.; Kasugai, Y.; Ooi, M.; Sakai, K.; Teshigawara, M.; Hasegawa, S.; et al. Design and application of NOBORU—NeutrOn Beam line for Observation and Research Use at J-PARC. *Nucl. Instrum. Methods Phys. Res. Sect. A* **2008**, *589*, 310. [CrossRef]

37. Shinohara, T.; Kai, T.; Oikawa, K.; Segawa, M.; Harada, M.; Nakatani, T.; Ooi, M.; Aizawa, K.; Sato, H.; Kamiyama, T.; et al. Final design of the Energy-Resolved Neutron Imaging System "RADEN" at J-PARC. *J. Phys. Conf. Ser.* **2016**, *746*, 012007. [CrossRef]

38. Seto, H.; Itoh, S.; Yokoo, T.; Endo, H.; Nakajima, K.; Shibata, K.; Kajimoto, R.; Ohira-Kawamura, S.; Nakamura, M.; Kawakita, Y.; et al. Inelastic and quasi-elastic neutron scattering spectrometers in J-PARC. *Biochim. Biophys. Acta Gen. Subj.* **2017**, *1861*, 3651–3660. [CrossRef] [PubMed]

39. Kajimoto, R.; Nakajima, K.; Nakamura, M.; Soyama, K.; Yokoo, T.; Oikawa, K.; Arai, M. Study of the neutron guide design of the 4SEASONS spectrometer at J-PARC. *Nucl. Instrum. Methods Phys. Res. Sect. A* **2009**, *600*, 185–188. [CrossRef]

40. Nakamura, M.; Kawakita, Y.; Kambara, W.; Aoyama, K.; Kajimoto, R.; Nakajima, K.; Ohira-Kawamura, S.; Ikeuchi, K.; Kikuchi, T.; Inamura, Y.; et al. Oscillating radial collimators for the chopper spectrometers at MLF in J-PARC. *JPS Conf. Proc.* **2015**, *8*, 036011. [CrossRef]

41. Iida, K.; Ikeuchi, K.; Ishikado, M.; Suzuki, J.; Kajimoto, R.; Nakamura, M.; Inamura, Y.; Arai, M. Energy- and Q-resolution investigations of a chopper spectrometer 4SEASONS at J-PARC. *JPS Conf. Proc.* **2014**, *1*, 014016. [CrossRef]

42. Mezei, F. The raison d'être of long pulse spallation sources. *J. Neutron Res.* **1997**, *6*, 3–32. [CrossRef]

43. Mezei, F.; Russina, M.; Schorr, S. The multiwavelength cold neutron time-of-flight spectrometer project IN500 at LANSCE. *Physica B* **2000**, *276–278*, 128–129. [CrossRef]

44. Iimura, S.; Matsuishi, S.; Miyakawa, M.; Taniguchi, T.; Suzuki, K.; Usui, H.; Kuroki, K.; Kajimoto, R.; Nakamura, M.; Inamura, Y.; et al. Switching of intra-orbital spin excitations in electron-doped iron pnictide superconductors. *Phys. Rev. B* **2013**, *88*, 060501(R). [CrossRef]

45. Ishii, K.; Fujita, M.; Sasaki, T.; Minola, M.; Dellea, G.; Mazzoli, C.; Kummer, K.; Ghiringhelli, G.; Braicovich, L.; Tohyama, T.; et al. High-energy spin and charge excitations in electron-doped copper oxide superconductors. *Nat. Commun.* **2014**, *5*, 3714. [CrossRef] [PubMed]

46. Wang, Q.; Shen, Y.; Pan, B.; Zhang, X.; Ikeuchi, K.; Iida, K.; Christianson, A.D.; Walker, H.C.; Adroja, D.T.; Abdel-Hafiez, M.; et al. Magnetic ground state of FeSe. *Nat. Commun.* **2016**, *7*, 12182. [CrossRef] [PubMed]

47. Hu, D.; Yin, Z.; Zhang, W.; Ewings, R.A.; Ikeuchi, K.; Nakamura, M.; Roessli, B.; Wei, Y.; Zhao, L.; Chen, G.; et al. Spin excitations in optimally P-doped BaFe$_2$(As$_{0.7}$P$_{0.3}$)$_2$ superconductor. *Phys. Rev. B* **2016**, *94*, 094504. [CrossRef]

48. Horigane, K.; Kihou, K.; Fujita, K.; Kajimoto, R.; Ikeuchi, K.; Ji, S.; Akimitsu, J.; Lee, C.H. Spin excitations in hole-overdoped iron-based superconductors. *Sci. Rep.* **2016**, *6*, 33303. [CrossRef] [PubMed]

49. Matsuura, M.; Kawamura, S.; Fujita, M.; Kajimoto, R.; Yamada, K. Development of spin-wave-like dispersive excitations below the pseudogap temperature in the high-temperature superconductor La$_{2-x}$Sr$_x$CuO$_4$. *Phys. Rev. B* **2017**, *95*, 024504. [CrossRef]

50. Itoh, S.; Yokoo, T.; Masuda, T.; Yoshizawa, H.; Soda, M.; Ikeda, Y.; Kawana, D.; Sato, T.J.; Nambu, Y.; Kuwahara, K.; et al. Science from the initial operation of HRC. *JPS Conf. Proc.* **2015**, *8*, 034001. [CrossRef]

51. Cooke, J.F.; Blackman, J.A. Calculation of neutron cross sections for interband transitions in semiconductors. *Phys. Rev. B* **1982**, *26*, 4410–4420. [CrossRef]

52. Passell, L.; Dietrich, O.W.; Als-Nielsen, J. Neutron scattering from the Heisenberg ferromagnets EuO and EuS. I. The exchange interactions. *Phys. Rev. B* **1976**, *14*, 4897–4907. [CrossRef]

53. Ishikawa, Y.; Yamada, K.; Taijma, K.; Fukamachi, K. Spin dynamics in the amorphous invar alloy of $Fe_{86}B_{14}$. *J. Phys. Soc. Jpn.* **1981**, *50*, 1958–1963. [CrossRef]

54. Robinson, R.A. Neutron Brillouin scattering with chopper spectrometers. *Physica B* **1989**, *156–157*, 557–560. [CrossRef]

55. Aisa, D.; Aisa, S.; Babucci, E.; Barocchi, F.; Cunsolo, A.; De Francesco, A.; Formisano, F.; Gahl, T.; Guarini, E.; Laloni, A.; et al. BRISP: A new thermal-neutron spectrometer for small-angle studies of disordered matter. *J. Non-Cryst. Solids* **2006**, *352*, 5130–5135. [CrossRef]

56. Fang, Z.; Nagaosa, N.; Takahashi, K.S.; Asamitsu, A.; Mathieu, R.; Ogasawara, T.; Yamada, H.; Kawasaki, M.; Tokura, Y.; Terakura, K. The Anomalous Hall effect and magnetic monopoles in momentum space. *Science* **2003**, *302*, 92–95. [CrossRef] [PubMed]

57. Itoh, S.; Endoh, Y.; Yokoo, T.; Ibuka, S.; Park, J.-G.; Kaneko, Y.; Takahashi, K.S.; Tokura, Y.; Nagaosa, N. Weyl fermions and spin dynamics of metallic ferromagnet $SrRuO_3$. *Nat. Commun.* **2016**, *7*, 11788. [CrossRef] [PubMed]

58. Nakajima, K.; Nakamura, M.; Kajimoto, R.; Osakabe, T.; Kakurai, K.; Matsuda, M.; Metoki, N.; Wakimoto, S.; Sato, T.J.; Itoh, S.; et al. Cold-neutron disk-chopper spectrometer at J-PARC. *J. Neutron Res.* **2007**, *15*, 13–21. [CrossRef]

59. Nakamura, M.; Nakajima, K.; Kajimoto, R.; Arai, M. Utilization of multiple incident energies on Cold-Neutron Disk-Chopper Spectrometer at J-PARC. *J. Neutron Res.* **2007**, *15*, 31–37. [CrossRef]

60. Mezei, F. Multiplexing chopper systems for pulsed neutron sources: Practical basics. *J. Phys. Soc. Jpn.* **2013**, *82*, SA025. [CrossRef]

61. Kajimoto, R.; Nakamura, M.; Osakabe, T.; Sato, T.J.; Nakajima, K.; Arai, M. Study of converging neutron guides for the Cold Neutron Double-Chopper Spectrometer at J-PARC. *Physica B* **2006**, *385–386*, 1236–1239. [CrossRef]

62. Kajimoto, R.; Nakajima, K.; Nakamura, M.; Osakabe, T.; Sato, T.J.; Arai, M. Curved neutron guide of the cold neutron disk-chopper spectrometer at J-PARC. *J. Neutron Res.* **2008**, *16*, 81–86. [CrossRef]

63. Nakajima, K.; Ohira-Kawamura, S.; Kikuchi, T.; Kajimoto, R.; Takahashi, N.; Nakamura, M.; Soyama, K.; Osakabe, T. Beam-transport optimization for cold-neutron spectrometer. *EPJ Web Conf.* **2015**, *83*, 03011. [CrossRef]

64. Kikuchi, T.; Nakajima, K.; Ohira-Kawamura, S.; Inamura, Y.; Yamamuro, O.; Kofu, M.; Kawakita, Y.; Suzuya, K.; Nakamura, M.; Arai, M. Mode-distribution analysis of quasielastic neutron scattering and application to liquid water. *Phys. Rev. E* **2013**, *87*, 062314. [CrossRef] [PubMed]

65. *Neutron Spin Echo, Lecture Notes in Physics Volume 128*; Mezei, F., Ed.; Springer: Berlin, Germany, 1982.

66. Gähler, R.; Golub, R. Neutron resonance spin echo, bootstrap method for increasing the effective magnetic field. *J. Phys. France* **1988**, *49*, 1195–1202. [CrossRef]

67. Richter, D.; Monkenbusch, M.; Arbe, A.; Colmenero, J. Neutron spin echo in polymer systems. *Adv. Polym. Sci.* **2005**, *174*, 1–221.

68. Nambu, Y.; Gardner, J.S.; MacLaughlin, D.E.; Stock, C.; Endo, H.; Jonas, S.; Sato, T.J.; Nakatsuji, S.; Broholm, C. Spin Fluctuations from Hertz to Terahertz on a Triangular Lattice. *Phys. Rev. Lett.* **2015**, *115*, 127202. [CrossRef] [PubMed]

69. Pappas, C.; Lelievre-Berna, E.; Falus, P.; Bentley, P.M.; Moskvin, E.; Grigoriev, S.; Fouquet, P.; Farago, B. Chiral Paramagnetic Skyrmion-like Phase in MnSi. *Phys. Rev. Lett.* **2009**, *102*, 197202. [CrossRef] [PubMed]

70. Kindervater, J.; Martin, N.; Haeussler, W.; Krautloher, M.; Fuchs, C.; Muehlbauer, S.; Lim, J.A.; Blackburn, E.; Boeni, P.; Peiderer, C. Neutron spin echo spectroscopy under 17 T magnetic field at RESEDA. *EPJ Web Conf.* **2015**, *83*, 03008. [CrossRef]

71. Oda, T. Study on Time-of-Flight Neutron Resonance Neutron Spin Echo Technique at J-PARC MLF BL06. Ph.D. Thesis, Kyoto University, Kyoto, Japan, 23 March 2016.

72. Oda, T.; Hino, M.; Kitaguchi, M.; Geltenbort, P.; Kawabata, Y. Pulsed neutron time-dependent intensity modulation for quasi-elastic neutron scattering spectroscopy. *Rev. Sci. Instrum.* **2016**, *87*, 105124. [CrossRef] [PubMed]

73. Matsuo, H.; Noguchi, Y.; Miyayama, M.; Suzuki, M.; Watanabe, A.; Sasabe, S.; Ozaki, T.; Mori, S.; Torii, S.; Kamiyama, T. Structural and piezoelectric properties of high-density $(Bi_{0.5}K_{0.5})TiO_3$-$BiFeO_3$ ceramics. *J. Appl. Phys.* **2010**, *108*, 104103. [CrossRef]

74. Park, J.; Lee, S.; Kang, M.; Jang, K.-H.; Lee, C.; Streltsov, S.; Mazurenko, V.; Valentyuk, M.; Medvedeva, J.; Kamiyama, T.; et al. Doping dependence of spin-lattice coupling and two-dimensional ordering in multiferroic hexagonal $Y_{1-x}Lu_xMnO_3$ ($0 \leq x \leq 1$). *Phys. Rev. B* **2010**, *82*, 054428. [CrossRef]

75. Kamaya, N.; Homma, K.; Yamakawa, Y.; Hirayama, M.; Kanno, R.; Yonemura, M.; Kamiyama, T.; Kato, Y.; Hama, S.; Kawamoto, K.; et al. A lithium superionic conductor. *Nat. Mater.* **2011**, *10*, 682–686. [CrossRef] [PubMed]

76. Takai, S.; Doi, Y.; Torii, S.; Zhang, J.; Surya Panca Putra, T.Y.; Miao, P.; Kamiyama, T.; Esaka, T. Structural and electrical properties of Pb-substituted $La_2Mo_2O_9$ oxide ion conductors. *Solid State Ionics* **2013**, *238*, 36–43. [CrossRef]

77. Lee, S.; Zhang, J.R.; Torii, S.; Choi, S.; Cho, D.-Y.; Kamiyama, T.; Yu, J.; McEwen, K.A.; Park, J.-G. Large in-plane deformation of RuO_6 octahedron and ferromagnetism of bulk $SrRuO_3$. *J. Phys. Condens. Matter* **2013**, *25*, 465601. [CrossRef] [PubMed]

78. Qasim, I.; Blanchard, P.E.R.; Kennedy, B.J.; Kamiyama, T.; Miao, P.; Torii, S. Structural and electronic properties of $Sr_{1-x}Ca_xTi_{0.5}Mn_{0.5}O_3$. *J. Solid State Chem.* **2014**, *213*, 293–300. [CrossRef]

79. Cheon, C.I.; Joo, H.W.; Chae, K.W.; Kim, J.S.; Lee, S.H.; Torii, S.; Kamiyama, T. Monoclinic ferroelectric $NaNbO_3$ at room temperature: Crystal structure solved by using super high resolution neutron powder diffraction. *Mater. Lett.* **2015**, *156*, 214–219. [CrossRef]

80. Gubkin, A.F.; Proskurina, E.P.; Kousaka, Y.; Sherokalova, E.M.; Selezneva, N.V.; Miao, P.; Lee, S.; Zhang, J.; Ishikawa, Y.; Torii, S.; et al. Crystal and magnetic structures of $Cr_{1/3}NbSe_2$ from neutron diffraction. *J. Appl. Phys.* **2016**, *119*, 013903. [CrossRef]

81. Torii, S.; Miao, P.; Lee, S.; Yonemura, M.; Kamiyama, T. *MLF Annual Report 2014*; Tokai, Japan, 2014; pp. 88–89.

82. Kino, K.; Mori, K.; Miyayama, M.; Yonemura, M.; Torii, S.; Kawai, M.; Fukunaga, T.; Kamiyama, T. Design of Air Scattering Chamber for the Powder Diffractometer SPICA. *J. Phys. Soc. Jpn.* **2011**, *80* (Suppl. B), SB001. [CrossRef]

83. Oishi, R.; Yonemura, M.; Nishimaki, Y.; Torii, S.; Hoshikawa, A.; Ishigaki, T.; Morishima, T.; Mori, K.; Kamiyama, T. Rietveld analysis software for J-PARC. *Nucl. Instrum. Methods Phys. Res. Sect. A* **2009**, *600*, 94–96. [CrossRef]

84. Taminato, S.; Yonemura, M.; Shiotani, S.; Kamiyama, T.; Torii, S.; Nagao, M.; Ishikawa, Y.; Mori, K.; Fukunaga, T.; Onodera, Y.; et al. Real-time observations of lithium battery reactions—Operando neutron diffraction analysis during practical operation. *Sci. Rep.* **2016**, *6*, 28843. [CrossRef] [PubMed]

85. Hoshikawa, A.; Ishigaki, T.; Yonemura, M.; Iwase, K.; Oguro, H.; Sulistyanintyas, D.; Kamiyama, T.; Hayashi, M. Automatic sample changer for IBARAKI materials design diffractometer (iMATERIA). *J. Phys. Conf. Ser.* **2010**, *251*, 012083. [CrossRef]

86. Sano-Furukawa, A.; Hattori, T.; Arima, H.; Yamada, A.; Tabata, S.; Kondo, M.; Nakamura, A.; Kagi, H.; Yagi, T. Six-axis multi-anvil press for high-pressure, high-temperature neutron diffraction experiments. *Rev. Sci. Instrum.* **2014**, *85*, 113905. [CrossRef] [PubMed]

87. Machida, A.; Saitoh, H.; Sugimoto, H.; Hattori, T.; Sano-Furukawa, A.; Endo, N.; Katayama, Y.; Iizuka, R.; Sato, T.; Mastuo, M.; et al. Site occupancy of interstitial deuterium atoms in face-centred cubic iron. *Nat. Commun.* **2014**, *5*, 5063. [CrossRef] [PubMed]

88. Komatsu, K.; Moriyama, M.; Koizumi, T.; Nakayama, K.; Kagi, H.; Abe, J.; Harjo, S. Development of a new P-T controlling system for neutron-scattering experiments. *High Pressure Res.* **2013**, *33*, 208. [CrossRef]

89. Komatsu, K.; Noritake, F.; Machida, S.; Sano-Furukawa, A.; Hattori, T.; Yamane, R.; Kagi, H. Partially ordered state of ice XV. *Sci. Rep.* **2016**, *6*, 28920. [CrossRef] [PubMed]

90. Klotz, S.; Komatsu, K.; Pietrucci, F.; Kagi, H.; Ludl, A.-A.; Machida, S.; Hattori, T.; Sano-Furukawa, A.; Bove, L.E. Ice VII from aqueous salt solutions: From a glass to a crystal with broken H-bonds. *Sci. Rep.* **2016**, *6*, 32040. [CrossRef] [PubMed]

91. Hemmi, T.; Harjo, S.; Nunoya, Y.; Kajitani, H.; Koizumi, N.; Aizawa, K.; Machiya, S.; Osamura, K. Neutron diffraction measurement of internal strain in the first Japanese ITER CS conductor sample. *Supercon. Sci. Tech.* **2013**, *26*, 084002. [CrossRef]

92. Suzuki, H.; Kusunoki, K.; Hatanaka, Y.; Mukai, T.; Tasai, A.; Kanematsu, M.; Kabayama, K.; Harjo, S. Measuring strain and stress distributions along rebar embedded in concrete using time-of-flight neutron diffraction. *Meas. Sci. Tech.* **2014**, *25*, 025602. [CrossRef]

93. Asoo, K.; Tomota, Y.; Harjo, S.; Okitsu, Y. Tensile behavior of a TRIP-aided ultra-fine grained steel studied by neutron diffraction. *ISIJ Int.* **2011**, *51*, 145–150. [CrossRef]

94. Zhang, Q.H.; Zhai, Z.; Nie, Z.H.; Harjo, S.; Cong, D.Y.; Wang, M.G.; Li, J.; Wang, Y.D. An in situ neutron diffraction study of anomalous superelasticity in a strain glass Ni43Fe18Ga27Co12 alloy. *J. Appl. Cryst.* **2015**, *48*, 1183–1191. [CrossRef]

95. Takahashi, K.; Oguro, H.; Awaji, S.; Watanabe, K.; Harjo, S.; Aizawa, K.; Machiya, S.; Suzuki, H.; Osamura, K. Prebending effect on three-dimensional strain in CuNb/(Nb,Ti)$_3$Sn wires under a tensile load. *IEEE Trans. Appl. Supercon.* **2012**, *22*, 6000204. [CrossRef]

96. Yamaguchi, T.; Fukuda, T.; Kakeshita, T.; Harjo, S.; Nakamoto, T. Neutron diffraction study on very high elastic strain of 6% in an Fe$_3$Pt under compressive stress. *Appl. Phys. Lett.* **2014**, *104*, 231908. [CrossRef]

97. Shi, Z.M.; Gong, W.; Tomota, Y.; Harjo, S.; Li, J.; Chi, B.; Pu, J. Study of tempering behavior of lath martensite using in situ neutron diffraction. *Mater. Charact.* **2015**, *107*, 29–32. [CrossRef]

98. Gong, W.; Tomota, Y.; Harjo, S.; Su, Y.H.; Aizawa, K. Effect of prior martensite on bainite transformation in nanobainite steel. *Acta Mater.* **2015**, *85*, 243–249. [CrossRef]

99. Ungár, T.; Harjo, S.; Kawasaki, T.; Tomota, Y.; Ribárik, G.; Shi, Z. Composite behavior of lath martensite steels induced by plastic strain, a new paradigm for the elastic-plastic response of martensitic steels. *Metal. Mater. Trans. A* **2017**, *48*, 159–167. [CrossRef]

100. Ohhara, T.; Kusaka, K.; Hosoya, T.; Kurihara, K.; Tomoyori, K.; Niimura, N.; Tanaka, I.; Suzuki, J.; Nakatani, T.; Otomo, T.; et al. Development of data processing software for a new TOF single crystal neutron diffractometer at J-PARC. *Nucl. Instrum. Methods Phys. Res. Sect. A* **2009**, *600*, 195–197. [CrossRef]

101. Yano, N.; Yamada, T.; Hosoya, T.; Ohhara, T.; Tanaka, I.; Kusaka, K. Application of profile fitting method to neutron time-of-flight protein single crystal diffraction data collected at the iBIX. *Sci. Rep.* **2016**, *6*, 36628. [CrossRef] [PubMed]

102. Yamada, N.L.; Mitamura, K.; Sagehashi, H.; Torikai, N.; Sato, S.; Seto, H.; Furusaka, M.; Oda, T.; Hino, M.; Fujiwara, T.; et al. Development of Sample Environments for the SOFIA Reflectometer for Seconds-Order Time-Slicing Measurements. *JPS Conf. Proc.* **2015**, *8*, 036003. [CrossRef]

103. Yonemura, M.; Hirayama, M.; Suzuki, K.; Kanno, R.; Torikai, N.; Yamada, N.L. Development of Spectroelectrochemical Cells for in situ Neutron Reflectometry. *IOP Conf. Ser.* **2014**, *502*, 012054. [CrossRef]

104. Machida, A.; Honda, M.; Hattori, T.; Sano-Furukawa, A.; Watanuki, T.; Katayama, Y.; Aoki, K.; Komatsu, K.; Arima, H.; Ohshita, H.; et al. Formation of NaCl-type monodeuteride LaD by the disproportionation reaction of LaD$_2$. *Phys. Rev. Lett.* **2012**, *108*, 205501. [CrossRef] [PubMed]

105. Takagi, S.; Iijima, Y.; Sato, T.; Saitoh, H.; Ikeda, K.; Otomo, T.; Miwa, K.; Ikeshoji, T.; Orimo, S.-I. Formation of novel transition metal hydride complexes with ninefold hydrogen coordination. *Sci. Rep.* **2017**, *7*, 44253. [CrossRef] [PubMed]

106. Mori, K.; Kasai, T.; Iwase, K.; Fujisaki, F.; Onodera, Y.; Fukunaga, T. Structural origin of massive improvement in Li-ion conductivity on transition from (Li$_2$S)$_5$(GeS$_2$)(P$_2$S$_5$) glass to Li$_{10}$GeP$_2$S$_{12}$ crystal. *Solid State Ionics* **2017**, *301*, 163–169. [CrossRef]

107. Saito, S.; Watanabe, H.; Hayashi, Y.; Matsugami, M.; Tsuzuki, S.; Seki, S.; Canongia Lopes, J.N.; Atkin, R.; Ueno, K.; Dokko, K.; et al. Li(+) Local Structure in Li-Tetraglyme Solvate Ionic Liquid Revealed by Neutron Total Scattering Experiments with the (6/7)Li Isotopic Substitution Technique. *J. Phys. Chem. Lett.* **2016**, *7*, 2832–2837. [CrossRef] [PubMed]

108. Hiraishi, M.; Iimura, S.; Kojima, K.M.; Yamaura, J.; Hiraka, H.; Ikeda, K.; Miao, P.; Ishikawa, Y.; Torii, S.; Miyazaki, M.; et al. Bipartite magnetic parent phases in the iron oxypnictide superconductor. *Nat. Phys.* **2014**, *10*, 300–303. [CrossRef]

109. Kin, T.; Furutaka, K.; Goko, S.; Harada, H.; Kimura, A.; Kitatani, F.; Nakamura, S.; Ohta, M.; Oshima, M.; Toh, Y.; et al. The "4π Ge Spectrometer" for Measurements of Neutron Capture Cross Sections by the TOF Method at the J-PARC/MLF/ANNRI. *J. Korean Phys. Soc.* **2011**, *59*, 1769–1772. [CrossRef]

110. Toh, Y.; Ebihara, M.; Kimura, A.; Nakamura, S.; Harada, H.; Hara, K.Y.; Koizumi, M.; Kitatani, F.; Furutaka, K. Synergistic Effect of Combining Two Nondestructive Analytical Methods for Multielemental Analysis. *Anal. Chem.* **2014**, *86*, 12030–12036. [CrossRef] [PubMed]

111. Kimura, A.; Fujii, T.; Fukutani, S.; Furutaka, K.; Goko, S.; Hara, K.Y.; Harada, H.; Hirose, K.; Hori, J.; Igashira, M.; et al. Neutron-capture cross-sections of ^{244}Cm and^{246}Cm measured with an array of large germanium detectors in the ANNRI at J-PARC/MLF. *J. Nucl. Sci. Technol.* **2012**, *49*, 708–724. [CrossRef]

112. Harada, H.; Ohta, M.; Kimura, A.; Furutaka, K.; Hirose, K.; Hara, K.Y.; Kin, T.; Kitatani, F.; Koizumi, M.; Nakamura, S.; et al. Capture Cross-section Measurement of ^{241}Am(n,γ) at J-PARC/MLF/ANNRI. *Nucl. Data Sheets* **2014**, *119*, 61–64. [CrossRef]

113. Hirose, K.; Furutaka, K.; Hara, K.Y.; Harada, H.; Kimura, A.; Kin, T.; Kitatani, F.; Koizumi, M.; Nakamura, S.; Oshima, M.; et al. Cross-section measurement of ^{237}Np (n, γ) from 10 meV to 1 keV at Japan Proton Accelerator Research Complex. *J. Nucl. Sci. Technol.* **2013**, *50*, 188–200. [CrossRef]

114. Kimura, A.; Hirose, K.; Nakamura, S.; Harada, H.; Hara, K.Y.; Hori, J.; Igashira, M.; Kamiyama, T.; Katabuchi, T.; Kino, K.; et al. Measurements of Neutron Capture Cross Sections of ^{112}Sn and ^{118}Sn with J-PARC/MLF/ANNRI. *Nucl. Data Sheets* **2014**, *119*, 150–153. [CrossRef]

115. Nakamura, S.; Kimura, A.; Kitatani, F.; Ohta, M.; Furutaka, K.; Goko, S.; Hara, K.Y.; Harada, H.; Hirose, K.; Kin, T.; et al. Cross Section Measurements of the Radioactive ^{107}Pd and STable 105,108Pd Nuclei at J-PARC/MLF/ANNRI. *Nucl. Data Sheets* **2014**, *119*, 143–146. [CrossRef]

116. Katabuchi, T.; Matsuhashi, T.; Terada, K.; Igashira, M.; Mizumoto, M.; Hirose, K.; Kimura, A.; Iwamoto, N.; Hara, K.Y.; Harada, H.; et al. Misassigned neutron resonances of ^{142}Nd and stellar neutron capture cross sections. *Phys. Rev. C* **2015**, *91*, 037603. [CrossRef]

117. Mishima, K. Neutron network news. *Hamon* **2015**, *25*, 156.

118. Imajo, S.; Mishima, K.; Kitaguchi, M.; Iwashia, Y.; Yamada, N.L.; Hino, M.; Oda, T.; Ino, T.; Shimizu, H.M.; Yamashita, S.; et al. Pulsed UCN Production using a Doppler Shifter at J-PARC. *Prog. Theor. Exp. Phys.* **2016**, *2016*, 013C02. [CrossRef]

119. Maekawa, F.; Harada, M.; Oikawa, K.; Teshigawara, M.; Kai, T.; Meigo, S.; Ooi, M.; Sakamoto, S.; Takada, H.; Futakawa, M.; et al. First neutron production utilizing J-PARC pulsed spallation neutron source JSNS and neutronic performance demonstrated. *Nucl. Instrum. Methods Phys. Res. Sect. A* **2010**, *620*, 159. [CrossRef]

120. Nakamura, M.; Kambara, W.; Krist, T.; Shinohara, T.; Ikeuchi, K.; Arai, M.; Kajimoto, R.; Nakajima, K.; Tanaka, H.; Suzuki, J.; et al. Feasibility demonstration of a new Fermi chopper with supermirror-coated slit package. *Nucl. Instrum. Methods Phys. Res. Sect. A* **2014**, *737*, 142. [CrossRef]

121. Nakamura, T.; Kawasaki, T.; Hosoya, T.; Toh, K.; Oikawa, K.; Sakasai, K.; Ebine, M.; Birumachi, A.; Soyama, K.; Katagiri, M. A large-area two-dimensional scintillator detector with a wavelength-shifting fibre readout for a time-of-flight single-crystal neutron diffractometer. *Nucl. Instrum. Methods Phys. Res. Sect. A* **2012**, *686*, 64. [CrossRef]

122. Kawasaki, T.; Nakamura, T.; Toh, K.; Hosoya, T.; Oikawa, K.; Ohhara, T.; Kiyanagi, R.; Ebine, M.; Birumachi, A.; Sakasai, K.; et al. Detector system of the SENJU single-crystal time-of-flight neutron diffractometer at J-PARC/MLF. *Nucl. Instrum. Methods Phys. Res. Sect. A* **2014**, *735*, 444. [CrossRef]

123. Nakamura, T.; Toh, K.; Kawasaki, T.; Honda, K.; Suzuki, H.; Ebine, M.; Birumachi, A.; Sakasai, K.; Soyama, K.; Katagiri, M. A scintillator-based detector with sub-100-um spatial resolution comprising a fibre-optic taper with wavelength-shifting fibre readout for time-of-flight neutron imaging. *Nucl. Instrum. Methods Phys. Res. Sect. A* **2014**, *737*, 176–183. [CrossRef]

124. Parker, J.D.; Hattori, K.; Fujioka, H.; Harada, M.; Iwaki, S.; Kabuki, S.; Kishimoto, Y.; Kubo, H.; Kurosawa, S.; Miuchi, K.; et al. Neutron imaging detector based on the micro-pixel chamber. *Nucl. Instrum. Methods Phys. Res. Sect. A* **2013**, *697*, 23. [CrossRef]

125. Segawa, M.; Kai, T.; SaKai, T.; Ooi, M.; Kureta, M. Development of a high-speed camera system for neutron imaging at a pulsed neutron source. *Nucl. Instrum. Methods Phys. Res. Sect. A* **2013**, *697*, 77. [CrossRef]

126. Parker, J.D.; Harada, M.; Hattori, K.; Iwaki, S.; Kabuki, S.; Kishimoto, Y.; Kubo, H.; Kurosawa, S.; Matsuoka, Y.; Miuchi, K.; et al. Spatial resolution of a μPIC-based neutron imaging detector. *Nucl. Instrum. Methods Phys. Res. Sect. A* **2013**, *726*, 155. [CrossRef]

127. Segawa, M.; Ooi, M.; Kai, T.; Shinohara, T.; Kureta, M.; Sakamoto, K.; Imaki, T. Development of a pulsed neutron three-dimensional imaging system using a highly sensitive image-intensifier at J-PARC. *Nucl. Instrum. Methods Phys. Res. Sect. A* **2015**, *769*, 97. [CrossRef]

128. Hasemi, H.; Harada, M.; Kai, T.; Shinohara, T.; Ooi, M.; Sato, H.; Kino, K.; Segawa, M.; Kamiyama, T.; Kiyanagi, Y. Evaluation of nuclide density by neutron resonance transmission at the NOBORU instrument in J-PARC/MLF. *Nucl. Instrum. Methods Phys. Res. Sect. A* **2015**, *773*, 137. [CrossRef]

129. Tremsin, A.S.; Shinohara, T.; Kai, T.; Ooi, M.; Kamiyama, T.; Kiyanagi, Y.; Shiota, Y.; McPhate, J.B.; Vallerga, J.V.; Siegmund, O.H.W.; et al. Neutron resonance transmission spectroscopy with high spatial and energy resolution at the J-PARC pulsed neutron source. *Nucl. Instrum. Methods Phys. Res. Sect. A* **2014**, *746*, 47. [CrossRef]

130. Ohoyama, K.; Lee, S.; Yoshii, S.; Narumi, Y.; Morioka, T.; Nojiri, H.; Jeon, G.S.; Cheong, S.W.; Park, J.G. High Field Neutron Diffraction Studies on Metamagnetic Transition of Multiferroic $BiFeO_3$. *J. Phys. Soc. Jpn.* **2011**, *80*, 125001. [CrossRef]

131. Makowska, M.G.; Strobl, M.; Lauridsen, E.M.; Frandsen, H.L.; Tremsin, A.S.; Shinohara, T.; Kuhn, L.T. Phase Transition Mapping by Means of Neutron Imaging in SOFC Anode Supports during Reduction under Applied Stress. *ECS Trans.* **2015**, *68*, 1103. [CrossRef]

132. Tremsin, A.S.; Ganguly, S.; Meco, S.M.; Pardal, G.R.; Shinohara, T.; Feller, W.B. Investigation of dissimilar metal welds by energy-resolved neutron imaging. *J. Appl. Cryst.* **2016**, *49*, 1130. [CrossRef] [PubMed]

133. Tremsin, A.S.; Gao, Y.; Dial, L.C.; Grazzi, F.; Shinohara, T. Investigation of microstructure in additive manufactured Inconel 625 by spatially resolved neutron transmission spectroscopy. *Sci. Technol. Adv. Mater.* **2016**, *17*, 326. [CrossRef] [PubMed]

134. Sato, H.; Sato, T.; Shiota, Y.; Kamiyama, T.; Tremsin, A.S.; Ohnuma, M.; Kiyanagi, Y. Relation between Vickers Hardness and Bragg-Edge Broadening in Quenched Steel Rods Observed by Pulsed Neutron Transmission Imaging. *Mater. Trans.* **2015**, *56*, 1147. [CrossRef]

135. Su, Y.H.; Oikawa, K.; Harjo, S.; Shinohara, T.; Kai, T.; Harada, M.; Hiroi, K.; Zhang, S.Y.; Parker, J.D.; Sato, H.; et al. Time-of-flight neutron Bragg-edge transmission imaging of microstructures in bent steel plates. *Mater. Sci. Eng. A* **2016**, *675*, 19. [CrossRef]

136. Tremsin, A.S.; Rakovan, J.; Shinohara, T.; Kockelmann, W.; Losko, A.S.; Vogel, S.C. Non-Destructive Study of Bulk Crystallinity and Elemental Composition of Natural Gold Single Crystal Samples by Energy-Resolved Neutron Imaging. *Sci. Rep.* **2017**, *7*, 40759. [CrossRef] [PubMed]

137. Miyajima, S.; Shishido, H.; Narukami, Y.; Yoshioka, N.; Fujimaki, A.; Hidaka, M.; Oikawa, K.; Harada, M.; Oku, T.; Arai, M.; et al. Neutron flux spectrum revealed by Nb-based current-biased kinetic inductance detector with a ^{10}B conversion layer. *Nucl. Instrum. Methods Phys. Res. Sect. A* **2017**, *842*, 71. [CrossRef]

138. Shishido, H.; Miyajima, S.; Narukami, Y.; Oikawa, K.; Harada, M.; Oku, T.; Arai, M.; Hidaka, M.; Fujimaki, A.; Ishida, T. Neutron detection using a current biased kinetic inductance detector. *Appl. Phys. Lett.* **2015**, *107*, 23601. [CrossRef]

139. Kai, T.; Hiroi, K.; Su, Y.; Shinohara, T.; Parker, J.D.; Matsumoto, Y.; Hayashida, H.; Segawa, M.; Nakatani, T.; Oikawa, K.; et al. Reliability estimation of neutron resonance thermometry using tantalum and tungsten. *Phys. Procedia* **2017**, *88*, 306–313. [CrossRef]

140. Sadeghilaridjani, M.; Kato, K.; Shinohara, T.; Yashiro, W.; Momose, A.; Kato, H. High aspect ratio grating by isochronal imprinting of less viscous workable Gd-based metallic glass for neutron phase imaging. *Intermetallics* **2016**, *78*, 55. [CrossRef]

141. Seki, Y.; Shinohara, T.; Parker, J.D.; Yashiro, W.; Momose, A.; Kato, K.; Kato, H.; Sadeghilaridjani, M.; Otake, Y.; Kiyanagi, Y. Development of Multi-colored Neutron Talbot-Lau Interferometer with Absorption Grating Fabricated by Imprinting Method of Metallic Glass. *J. Phys. Soc. Jpn.* **2017**, *86*, 044001. [CrossRef]

142. Kanno, R.; Hata, T.; Kawamoto, Y.; Irie, M. Synthesis of a new lithium ionic conductor, thio-LISICON–lithium germanium sulfide system. *Solid State Ionics* **2000**, *130*, 97–104. [CrossRef]

143. Kanno, R.; Murayama, M. Lithium Ionic Conductor Thio-LISICON: The Li_2S-GeS_2-P_2S_5 System. *J. Electrochem. Soc.* **2001**, *148*, A742–A746. [CrossRef]

144. Murayama, M.; Kanno, R.; Kawamoto, Y.; Kamiyama, T. Structure of the thio-LISICON, Li_4GeS_4. *Solid State Ionics* **2002**, *154–155*, 789–794. [CrossRef]

145. Hong, H.Y.-P. Crystal Structure and Ionic Conductivity of $Li_{14}Zn(GeO_4)_4$ and Other New Li+ Superionic Conductors. *Mater. Res. Bull.* **1978**, *13*, 117–124. [CrossRef]

146. Kato, Y.; Hori, S.; Saito, T.; Suzuki, K.; Hirayama, M.; Mitsui, A.; Yonemura, M.; Iba, H.; Kanno, R. High-power all-solid-state batteries using sulfide superionic conductors. *Nat. Energy* **2016**, *1*, 16030. [CrossRef]

147. Ogo, S.; Kabe, R.; Uehara, K.; Kure, B.; Nishimura, T.; Menon, S.C.; Harada, R.; Fukuzumi, S.; Higuchi, Y.; Ohhara, T.; et al. A Dinuclear Ni(μ-H)Ru Complex Derived from H_2. *Science* **2007**, *316*, 585–587. [CrossRef] [PubMed]

148. Ogo, S.; Ichikawa, K.; Kishima, T.; Matsumoto, T.; Nakai, H.; Kusaka, K.; Ohhara, T. A Functional [NiFe]Hydrogenaze Mimic That Catalyzes Electron and Hydride Transfer from H_2. *Science* **2013**, *339*, 682–684. [CrossRef] [PubMed]

149. Kobayashi, M.; Terayama, Y.; Yamaguchi, H.; Terada, M.; Murakami, D.; Ishihara, K.; Takahara, A. Wettability and Antifouling Behavior on the Surfaces of Superhydrophilic Polymer Brushes. *Langmuir* **2012**, *28*, 7212. [CrossRef] [PubMed]

150. Morse, A.J.; Edmondson, S.; Dupin, D.; Armes, S.P.; Zhang, Z.; Leggett, G.J.; Thompson, R.L.; Lewis, A.L. Biocompatible polymer brushes grown from model quartz fibres: Synthesis, characterisation and in situ determination of frictional coefficient. *Soft Matter* **2010**, *6*, 1571. [CrossRef]

151. Kobayashi, M.; Terayama, Y.; Kikuchi, M.; Takahara, A. Chain dimensions and surface characterization of superhydrophilic polymer brushes with zwitterion side groups. *Soft Matter* **2013**, *9*, 5138. [CrossRef]

152. Umegaki, I.; Tanaka, H.; Kurita, N.; Ono, T.; Laver, M.; Niedermayer, C.; Rüegg, C.; Ohira-Kawamura, S.; Nakajima, K.; Kakurai, K. Spinon, soliton, and breather in the spin-1/2 antiferromagnetic chain compound $KCuGaF_6$. *Phys. Rev. B* **2015**, *92*, 174412. [CrossRef]

153. Khanh, N.D.; Abe, N.; Sagayama, H.; Nakao, A.; Hanashima, T.; Kiyanagi, R.; Tokunaga, Y.; Arima, T. Magnetoelectric coupling in the honeycomb antiferromagnet $Co_4Nb_2O_9$. *Phys. Rev. B* **2016**, *93*, 075117. [CrossRef]

154. Luo, H.; Lu, X.; Zhang, R.; Wang, M.; Goremychkin, E.A.; Adroja, D.T.; Danilkin, S.; Deng, G.; Yamani, Z.; Dai, P. Electron doping evolution of the magnetic excitations in $BaFe_{2-x}Ni_xAs_2$. *Phys. Rev. B* **2013**, *88*, 144516. [CrossRef]

155. Wang, M.; Zhang, C.; Lu, X.; Tan, G.; Luo, H.; Song, Y.; Wang, M.; Zhang, X.; Goremychkin, E.A.; Perring, T.G.; et al. Doping dependence of spin excitations and its correlations with high-temperature superconductivity in iron pnictides. *Nat. Commun.* **2013**, *4*, 2874. [CrossRef] [PubMed]

156. Lee, C.H.; Kihou, K.; Kawano-Furukawa, H.; Saito, T.; Iyo, A.; Eisaki, H.; Fukazawa, H.; Kohori, Y.; Suzuki, K.; Usui, H.; et al. Incommensurate spin fluctuations in hole-overdoped superconductor KFe_2As_2. *Phys. Rev. Lett.* **2011**, *106*, 067003. [CrossRef] [PubMed]

157. Knolle, J.; Eremin, I.; Chubukov, A.V.; Moessner, R. Theory of itinerant magnetic excitations in the spin-density-wave phase of iron-based superconductors. *Phys. Rev. B* **2010**, *81*, 140506(R). [CrossRef]

158. Kaneshita, E.; Tohyama, T. Spin and charge dynamics ruled by antiferromagnetic order in iron pnictide superconductors. *Phys. Rev. B* **2010**, *82*, 094441. [CrossRef]

159. Kovacic, M.; Christensen, M.H.; Gastiasoro, M.N.; Andersen, B.M. Spin excitations in the nematic phase and the metallic stripe spin-density wave phase of iron pnictides. *Phys. Rev. B* **2015**, *91*, 064424. [CrossRef]

160. Endoh, Y.; Böni, P. Magnetic excitations in metallic ferro- and antiferromagnets. *J. Phys. Soc. Jpn.* **2006**, *75*, 111002. [CrossRef]

161. Hayden, S.M.; Doubble, R.; Aeppli, G.; Perring, T.G.; Fawcett, E. Strongly enhanced magnetic excitations near the quantum critical point of $Cr_{1-x}V_x$ and why strong exchange enhancement need not imply heavy Fermion behavior. *Phys. Rev. Lett.* **2000**, *84*, 999. [CrossRef] [PubMed]

162. Tomiyoshi, S.; Yamaguchi, Y.; Ohashi, M.; Cowley, E.R.; Shirane, G. Magnetic excitations in the itinerant antiferromagnets Mn_3Si and Fe-doped Mn_3Si. *Phys. Rev. B* **1987**, *36*, 2181. [CrossRef]

163. Squires, G.L. *Introduction to the Theory of Thermal Neutron Scattering*; Cambridge University Press: Cambridge, UK, 1978.

quantum beam science

MDPI

Review

Materials and Life Science Experimental Facility at the Japan Proton Accelerator Research Complex III: Neutron Devices and Computational and Sample Environments

Kaoru Sakasai [1,*], Setsuo Satoh [2], Tomohiro Seya [2], Tatsuya Nakamura [1], Kentaro Toh [1], Hideshi Yamagishi [3], Kazuhiko Soyama [1], Dai Yamazaki [1], Ryuji Maruyama [1], Takayuki Oku [1], Takashi Ino [2], Hiroshi Kira [4], Hirotoshi Hayashida [4], Kenji Sakai [1], Shinichi Itoh [2], Kentaro Suzuya [1], Wataru Kambara [1], Ryoichi Kajimoto [1], Kenji Nakajima [1], Kaoru Shibata [1], Mitsutaka Nakamura [1], Toshiya Otomo [2], Takeshi Nakatani [1], Yasuhiro Inamura [1], Jiro Suzuki [5], Takayoshi Ito [4], Nobuo Okazaki [4], Kentaro Moriyama [4], Kazuya Aizawa [1], Seiko Ohira-Kawamura [1] and Masao Watanabe [1]

[1] Materials and Life Science Division, J-PARC Center, Japan Atomic Energy Agency (JAEA), Tokai, Ibaraki 319-1195, Japan; nakamura.tatsuya@jaea.go.jp (T.N.); toh.kentaro@jaea.go.jp (K.T.); soyama.kazuhiko@jaea.go.jp (K.So.); dai.yamazaki@j-parc.jp (D.Y.); ryuji.maruyama@j-parc.jp (R.M.); takayuki.oku@j-parc.jp (T.Ok.); kenji.sakai@j-parc.jp (Ke.S.); suzuya.kentaro@jaea.go.jp (K.Su.); kambara.wataru@jaea.go.jp (W.K.); ryoichi.kajimoto@j-parc.jp (R.K.); kenji.nakajima@j-parc.jp (K.N.); shibata.kaoru@jaea.go.jp (K.Sh.); nakamura.mitsutaka@jaea.go.jp (M.N.); takeshi.nakatani@j-parc.jp (T.N.); yasuhiro.inamura@j-parc.jp (Y.I.); aizawa.kazuya@jaea.go.jp (K.A.); seiko.kawamura@j-parc.jp (S.O.-K.); masao.watanabe@j-parc.jp (M.W.)

[2] Institute of Materials Structure Science, High Energy Accelerator Research Organization (KEK), Tsukuba, Ibaraki 300-3256, Japan; setsuo.satoh@kek.jp (S.S.); seyat@post.j-parc.jp (T.S.); takashi.ino@kek.jp (T.In.); shinichi.itoh@kek.jp (S.I.); toshiya.otomo@kek.jp (T.Ot.)

[3] Nippon Advanced Technology, Ltd., Tokai, Ibaraki 319-1106, Japan; yamagishi.hideshi@jaea.go.jp (H.Y.)

[4] Neutron Science and Technology Center, Comprehensive Research Organization for Science and Society (CROSS), Tokai, Ibaraki 319-1106, Japan; h_kira@cross.or.jp (H.K.); h_hayashida@cross.or.jp (H.H.); t_ito@cross.or.jp (T.It.); n_okazaki@cross.or.jp (N.O.); k_moriyama@cross.or.jp (K.M.)

[5] Computing Research Center, High Energy Accelerator Research Organization (KEK), Tsukuba, Ibaraki 300-3256, Japan; jiro.suzuki@j-parc.jp (J.S.)

* Correspondence: sakasai.kaoru@jaea.go.jp; Tel.: +81-29-284-3519

Received: 12 May 2017; Accepted: 24 July 2017; Published: 3 August 2017

Abstract: Neutron devices such as neutron detectors, optical devices including supermirror devices and ^{3}He neutron spin filters, and choppers are successfully developed and installed at the Materials Life Science Facility (MLF) of the Japan Proton Accelerator Research Complex (J-PARC), Tokai, Japan. Four software components of MLF computational environment, instrument control, data acquisition, data analysis, and a database, have been developed and equipped at MLF. MLF also provides a wide variety of sample environment options including high and low temperatures, high magnetic fields, and high pressures. This paper describes the current status of neutron devices, computational and sample environments at MLF.

Keywords: neutron detector; neutron supermirror; ^{3}He neutron spin filter; chopper; data acquisition; data analysis; database; sample environment

PACS: 29.85.Ca; 29.40.Cs; 29.40.Mc; 03.75.Be; 07.60.-j; 32.80.Bx; 29.25.Dz; 29.30.Hs; 07.05.Hd; 07.05.Kf; 07.20.Hy; 07.20.Mc; 07.55.Db

1. Introduction

Neutron devices play an important role in the work that goes on at neutron scattering facilities. At the Materials and Life Science Facility (MLF) of the Japan Proton Accelerator Research Complex (J-PARC), we have been developing advanced neutron devices such as neutron detectors, supermirror devices, ^{3}He neutron spin filters, and choppers with high performance at MLF. On the other hand, in a large facility such as MLF, sophisticated computational environment is indispensable for data acquisition (DAQ) and analysis. At MLF, four software components, instrument control, DAQ, data analysis, and a database, have been equipped. Furthermore, the sample environment (SE) is very important for neutron scattering experiments. Therefore, a special SE team is organized to operate, perform maintenance, and develop common SE equipment at MLF. In this paper, the state of the art of neutron devices, computational environment, and sample environment at MLF are described.

2. Neutron Devices

2.1. Neutron Detectors

2.1.1. A Neutron Encode with High Speed Network—(NeuNET) Module for ^{3}He Position Sensitive Detector

Several neutron instruments at MLF require large detector systems that cover very large solid angles and have a high pixel resolution and many time-of-flight (TOF) channels. More than 1000 linear position-sensitive ^{3}He gas detectors (PSDs) are used in instruments at MLF, such as high-resolution powder diffractometers and small-/wide-angle diffractometers. We have developed a new neutron encode with high speed network —(NeuNET) module and a large-scale DAQ system to satisfy the abovementioned requirements of large detector systems; the NeuNET module and DAQ system have an outstanding ability to process a large number of PSD signals. The NeuNET module can process data from PSDs having lengths in the range of 60 cm to 3 m, and it comprises a high-speed network (SiTCP) that lends flexibility and scalability to the DAQ system.

The NeuNET module combines the neutron measurement technology [1] developed at the KEK Neutron Science Laboratory (Tsukuba, Japan) and the network technology [2] developed at the KEK Institute of Particle and Nuclear Studies (Tsukuba, Japan). Figure 1a shows a picture of the NeuNET module. One NeuNET module that comprises one slot and whose height is double that specified by Versa module Europe (VME) standards processes the data of eight PSDs. Figure 1b shows a block diagram of the NeuNET module. When the PSD captures a neutron, the daughter board receives the signals of both sides (left and right) of the PSD, converting neutron signals into digital data via two analog-to-digital converters (ADCs). The field programmable gate array (FPGA) on the board detects the peak of the signal to calculate the pulse height from the difference between the peak and the baseline. If the sum of the pulse heights of the left and right signals exceeds the threshold, the signals are stored as the captured neutron data. The pulse heights and a time stamp at the detected peak are sent to the main FPGA on the main board.

The SiTCP system transfers the data from the NeuNET module to the DAQ system by means of a high-speed network without a CPU. The network is of the 100BASE-TX standard, and can transfer the data at almost the maximum speed (11 Mbyte/s) by the Transmission Control Protocol/Internet Protocol (TCP/IP). A standard PSD generates neutron data 30 k-cps (count per second) at most. Because the NeuNET module transmits the data of eight PSDs, the maximum data rate is 8 × 30 k-cps × 8 bytes = 1.42 M-byte/s. Therefore, the NeuNET module has a fairly high network capacity. The DAQ system with the NeuNET module is the de facto standard at MLF; it is used in more than half of the experimental spectrometers at this facility and can control thousands of ^{3}He gas detectors.

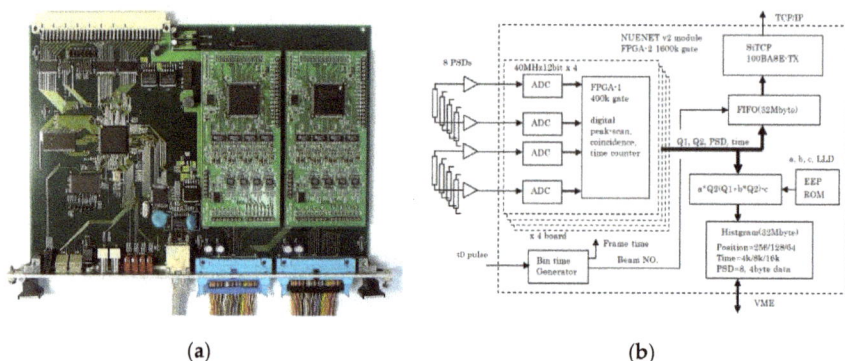

Figure 1. NeuNET: (**a**) Picture; (**b**) Block diagram. ADC: analog-to-digital converter; FPGA: field programmable gate array; PSD: linear position-sensitive ^{3}He gas detectors; TCP/IP: transmission control protocol / internet protocol; FIFO: first in first out memory; LLD: lower level discriminator; EEP ROM: electrical erasing programmable read only memory; VME: versa module Europa.

2.1.2. Scintillator Detectors

The Engineering Materials Diffractometer, TAKUMI at BL19, is one of the neutron instruments that are installed at the MLF. TAKUMI is specially designed for the analysis of residual stress and crystallographic structure of industrial materials. The commissioning of TAKUMI started in September 2008, and was completed in March 2009. The neutron detectors installed at TAKUMI [3,4] are large one-dimensional neutron detectors with a sensitive area of 1000 mm × 20 cm and a position resolution of 3 mm, as shown in Figure 2a, and have been developed under international collaboration with ISIS, Rutherford Appleton Laboratory (Didcot, UK).

The detectors were designed based on those installed in the ENGIN-X instrument of ISIS. The TAKUMI detector has 360 pixels with each size of 3 mm × 20 cm, whilst ENGIN-X detectors 240 pixels with a neutron sensitive area of 750 mm × 20 cm. TAKUMI has operated for more than 5 years with 10 detectors. In 2014–2015, 2 detectors of this type were fabricated and delivered to BL19, and thus a total of 12 detectors are now working at TAKUMI. These detectors have a neutron sensitivity of more than 50% at 1 Å and gamma-ray sensitivity of less than 10^{-6} at a gamma-ray energy of 1.3 MeV. The performance meets the requirements for residual stress analysis in neutron scattering experiments at TAKUMI.

The Single-crystal Neutron Diffractometer under Extreme Condition, SENJU, is a time-of-flight Laue single crystal diffractometer constructed at BL18. The aim of SENJU is to study crystal structures of materials under extreme environmental conditions, such as low temperature and high magnetic fields. We have developed a large-area scintillator detector using a scintillator and wavelength-shifting fiber technology [5,6]. To meet the detector specifications required for the SENJU, we designed a dedicated detector head that incorporates a 1-mm diameter wavelength-shifting fiber placed with a regular interval of 4 mm. Thicknesses of highly efficient ZnS/^{10}B$_2$O$_3$ scintillator screen were also optimized to ensure the detector efficiency. Figure 2b shows the developed detector module.

The detector exhibited a neutron-sensitive area of 256 × 256 mm^2, spatial resolution of 4 × 4 mm^2, and detection efficiency of 40% for 1.6-Å neutrons. The large-area detector system that includes 37 detector modules has been in service in the beamline since 2012.

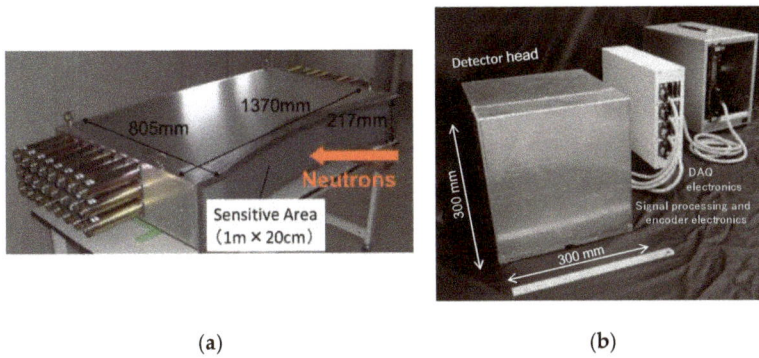

(a)   (b)

Figure 2. Photographs of (**a**) the neutron detector developed for TAKUMI and (**b**) the neutron detector developed for SENJU. DAQ: Data acquisition.

2.1.3. Gas-Based Two-Dimensional Detector

We have developed a ^{3}He gas-based two-dimensional neutron detector system for use in neutron scattering experiments using high-intensity pulsed neutrons at the Japan Proton Accelerator Research Complex (J-PARC) [7,8]. Figure 3 shows photographs of the developed detector system and of the detector head arranged in the beamline. This system exhibits superior performance, including a counting rate greater than several hundred k-cps, an average spatial resolution of less than 2.0 mm full-width at half-maximum (FWHM) with a standard deviation of 0.85 mm in the sensitive region, and a thermal neutron detection efficiency of 80%.

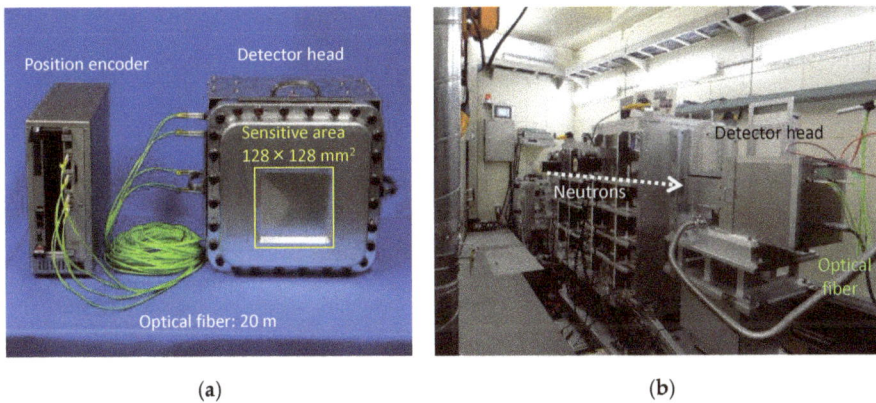

(a)   (b)

Figure 3. Photographs of (**a**) developed gas-based two-dimensional neutron detector system using individual line readout and optical signal transmission and (**b**) detector head in the beamline.

In general, a two-dimensional gaseous neutron detector has a few hundred signal lines along the vertical and horizontal axes, and the signal lines along each axis are usually connected together to conduct signal processing. The detector we have developed employs an individual line readout method, and the signal of each line is individually amplified, shaped, and discriminated by front-end electronics. A short response time and high spatial resolution can be obtained using this method. Although it is necessary to increase the gas pressure to achieve higher detection efficiency, increasing this pressure decreases the amplitude of the output neutron signal, and, consequently, discrimination between the neutron and background signals becomes difficult. Therefore, we developed a high-density multiwire-type detector element, a low-parallax pressure vessel with multichannel feedthroughs for neutron signals, multichannel front-end electronics using dedicated integrated circuits such as application specific integrated circuits and FPGA, and optical signal transmission devices specifically for our individual-readout detector system. These dedicated devices enabled the detector system to achieve high detection efficiency, short response time, and high spatial resolution. The optical signal transmission could also be used to establish long-distance transmission between the detector head and the data acquisition device without any electrical noise.

2.2. Supermirror Devices

A supermirror [9] is composed of a total reflection mirror and sequence of multilayer monochromators whose Bragg wavelength gradually changes. The development of high-performance neutron supermirrors is important for neutron experiments since it leads to a considerable increase in the available neutron intensity at the end of the guide and beam condensers such as focusing mirrors.

A multilayer with a smaller lattice-spacing (thickness of each layer) is desirable to extend the critical angle of the supermirror, which is needed in various applications. One of the most important problems in producing a small lattice-spacing multilayer is the interface roughness, which increases with an increasing number of bilayers. If the value of the root-mean-square (rms) interface roughness can be kept small compared with the lattice-spacing, high reflectivity can be achieved by stacking a sufficient number of layers. We have developed neutron supermirrors using the ion beam sputtering (IBS) technique because it has the advantage that target atoms are sputtered with higher energy compared with the other deposition techniques. It brings good quality of layers with higher density and small grain size. In addition, the layers are not easily peeled off due to the anchoring effect [10]. In this section, we summarize the development of high-performance supermirrors and its application to MLF.

2.2.1. Fabrication and Characterization

Supermirror Coating

An IBS system with an effective deposition area of 0.2 m^2 (diameter of 500 mm) has been installed at the Japan Atomic Energy Agency (JAEA). The difference in the deposition rate has been corrected to be less than 4% over the entire deposition area by using the mask just in front of the substrate holder. The pressure before and during the process is 1×10^{-5} and 2×10^{-2} Pa, respectively.

Ni/Ti supermirrors have been fabricated with a large critical angle extended to $m = 3, 4$, and 6.7, where m is the ratio of the critical angle of the supermirror to that of Ni. Neutron reflectivities of the supermirrors at the critical angle are 0.82, 0.66, and 0.23, respectively as shown in Figure 4a [10–12] (reprinted from [10], with permission from Elsevier). NiC/Ti [13,14] supermirror with $m = 4$ was fabricated using the same IBS system. The reflectivitiy of the supermirror at the critical angle was 0.82, as shown in Figure 4b (reprinted from [14], with permission from Elsevier). The test fabrication of the NiC/Ti supermirror with $m = 6$ was performed. A reflectivity of 0.40 was realized at a momentum transfer of 1.29 nm^{-1}, corresponding to $m = 6$ [14].

Neutrons scattered from a supermirror can be divided into specular and off-specular (diffuse) components. Suppression of the diffuse component is important since it creates a serious problem of low signal-to-noise ratio when it is used in a focusing system for such purposes as a small angle scattering measurement. The diffuse intensity was decreased by more than one order of magnitude by adopting the NiC/Ti supermirror instead of the conventional Ni/Ti supermirror as shown in Figure 5 [14] (Figure 5 is reprinted from [15], with permission from American Institute of Physics). This result implies that a high-performance focusing system with a noise level down to the order of 10^{-5} compared with the focused intensity can be realized by using a NiC/Ti supermirror.

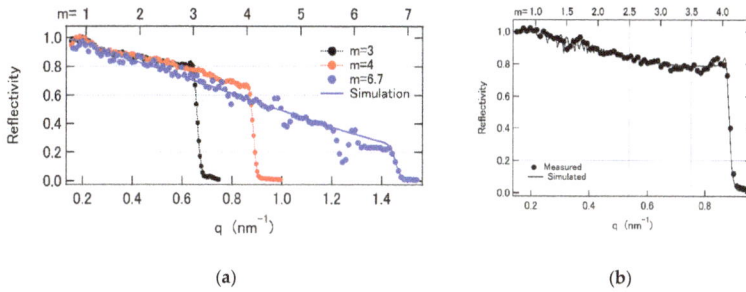

(a) (b)

Figure 4. (**a**) Measured and simulated reflectivities of Ni/Ti supermirror with m = 3, 4, and 6.7 (8001 layers). (**b**) Measured ($\circ$) and simulated (solid line) reflectivities of NiC/Ti supermirror with m = 4. Statistical errors are less than the size of the symbols.

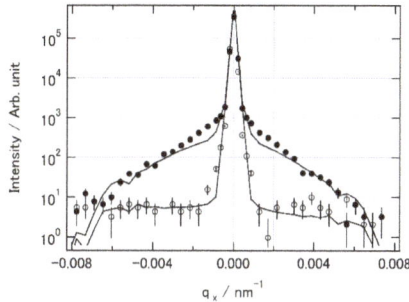

Figure 5. Measured diffuse intensity profiles of the Ni/Ti ($\bullet$) and NiC/Ti ($\circ$) supermirrors (m = 3). (a) Rocking scan with q_z = 0.62 nm^{-1}. Solid lines indicate the calculated profiles.

Supermirror Guide

For the production of neutron guides, supermirrors were coated on float glass and borosilicate float glass substrates for the inelastic neutron scattering instruments of 4D-Space Access Neutron Spectrometer (4SEASONS) at BL01, Biomolecular Dynamics Spectrometer (DNA) at BL02, and Cold-Neutron Disk-Chopper Spectrometer (AMATERAS) at BL14 [16,17]. Figure 6 shows the guide elements produced for these instruments. The typical value of the root mean square (rms) surface roughness of the substrate was 0.3 nm and the flatness was 3×10^{-4} rad. The elliptic curve of the guide was approximately realized by the tapered guide elements. To reduce the irradiation damage due to the neutron capture reaction, the use of the borosilicate glass was avoided at the guide section close to the moderator.

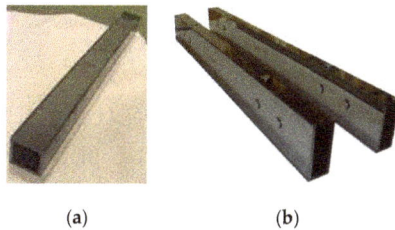

(a) (b)

Figure 6. (**a**) A guide element at the upper stream of BL01 with a square cross-section of 90×90 mm^2. (**b**) Guide elements of BL14 with a rectangular cross-section of 90×30 mm^2.

Focusing Mirror

For spallation neutron scattering facilities, achromatic focusing optics is particularly important because most experiments are performed with wideband and focal point beams, which are indispensable for measuring micrometer-scale samples, second-resolution time-resolved measurements, or high *q*-resolution small angle scattering measurements. Reflective focusing mirror is a prominent candidate for that purpose. A focusing mirror needs to accept and reflect wideband beam with large divergence into a small area with high efficiency and a high signal-to-noise ratio. In order to meet these demands, we have developed ultra-precise focusing mirrors combining a high-performance neutron supermirror and a very precise surface figuring technique. A mirror substrate was prepared by figuring a surface of synthesized quartz glass into a planoelliptical shape through the Numerically Controlled Local Wet Etching (NC-LWE) process [18]. NiC/Ti supermirror (*m* = 4) was deposited on the quartz glass over 90 × 40 mm. The focal length of the ellipsoid is 1050 mm. Wideband neutrons of λ > 3.64 Å were focused with focal spot size down to 0.25 mm and peak intensity gain of up to 6 without significant diffuse scattering, as shown in Figure 7 [19]. Time-of-flight measurements suggest that wideband neutrons are effectively focused to the focal point.

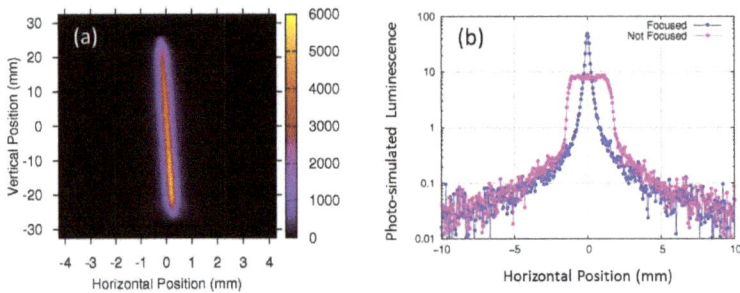

Figure 7. (**a**) Two-dimensional image of the focused beam. The tilt of the focused beam comes from the tilt of the image plate which was used in observation. (**b**) Horizontal intensity profiles of focused and unfocused beam at the vertical center.

We have developed a two-dimensional focusing device, a so-called Kirkpatrick–Baez (KB) mirror [20]. The KB mirror system is a quasi-aberration free system that consists of two total reflection elliptical mirrors placed at separate positions. One mirror with *m* = 4 and a length of 400 mm is used for vertical focusing and the other with *m* = 3 and a length of 100 mm is used for horizontal focusing. The focal length of the ellipsoid is 2100 mm. Elliptical quartz substrates were prepared by a fabrication process that combines conventional precision grinding, NC-LWE figuring and low-pressure polishing techniques. The NiC/Ti super-mirror was deposited using the IBS technique on the elliptical quartz surface. We obtained a figure error of less than 1 μm P–V (peak to valley) with a surface roughness of less than 0.3 nm rms on the supermirror. The measured focused beam width was 0.5 × 0.5 mm in FWHM, as shown in Figure 8.

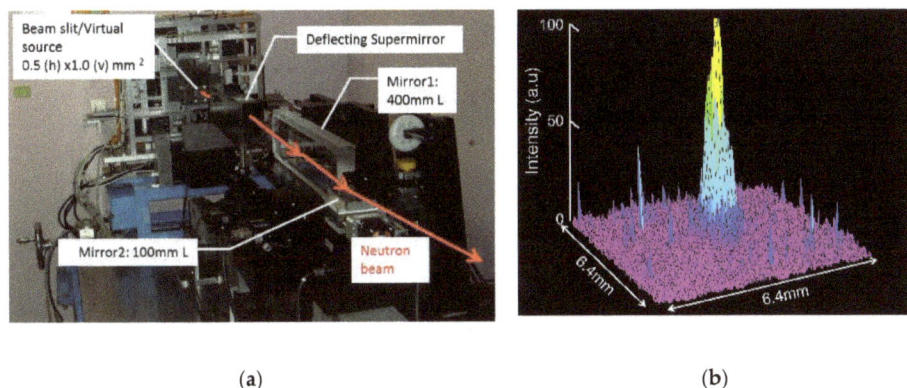

Figure 8. (**a**) KB mirror configuration at NOBORU of BL10. (**b**) Focused neutron beam by the Kirkpatrick–Baez (KB) mirror has a full-width at half-maximum (FWHM) of 0.5 mm.

2.3. ³He Neutron Spin Filters

In order to apply a ^{3}He neutron spin filter (NSF) to experiments at a pulsed neutron experimental facility, it is important to make the setup stable and easy to set up and operate, because the setup is located inside a radiation shield for high-energy gamma rays and neutrons. Moreover, a uniform magnetic field is essential to achieve a very high nuclear polarization. In this study, we have developed compact laser optics with a volume holographic grating (VHG) element and a flexible non-magnetic heater with a thickness of less than a half millimeter and a thermal tolerance up to 300 °C for a spin-exchange optical pumping (SEOP) method, and composed a setup for an in situ SEOP ^{3}He NSF. The design and performance of the setup will be described in the following sections.

2.3.1. Compact Laser Optics

We have developed compact laser optics with an air-cooled laser diode array (LDA) which does not need a water chiller to operate it. As the output laser power of the air-cooled LDA, which is around 30–40 W, is much lower than that of a water-cooled one, it is enough to polarize ^{3}He gas within a small-sized cell [21]. The laser spectrum of the LDA is broader than the absorption line width of Rb [22]. Thus, we employed a VHG element to create an external cavity laser (ECL) to narrow the width of the laser spectrum to match it to the absorption line width of Rb [23] (Figure 9). The developed laser optics are shown in Figure 9. The LDA is cooled with a Peltier device and a fan. The temperature of the VHG is controlled within ±1 °C using a heater. Collimating and shaping lenses and a λ/4 wave plate were assembled in a box (Figure 9). Therefore, the circularly polarized laser, which is necessary for the SEOP [21], is extracted from the box.

The measured laser spectra with and without the feedback are shown in Figure 10. The laser spectrum was narrowed to be a FWHM of 0.35 nm by the feedback, which is 16% of the original spectrum width without the feedback (Figure 10). The specification value of the spectrum width of the VHG element is 0.14 nm. The observed spectrum width obtained with the feedback was about 2.5 times that of the specification value of the VHG. This may be due to a miss-alignment of the optical components such as the so-called *smile* of the LDA (The *smile* is defined as the bending of the line of emitters of the LDA [24,25]). Here, the value of the *smile* of the LDA was unknown and was not also guaranteed by the supplier. To check for the *smile* effect, we tentatively set up the optics with the same geometry with a LDA which had the lower *smile* of less than 1 μm, and evaluated the laser spectrum. The obtained spectrum width was a FWHM of 0.18 nm (Figure 10), which was closer to the specification value of the VHG.

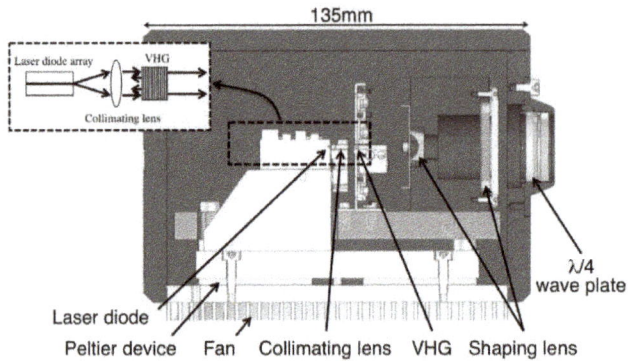

Figure 9. A drawing of the developed laser optics. The inset shows the schematic layout of the external cavity. VHG: volume holographic grating.

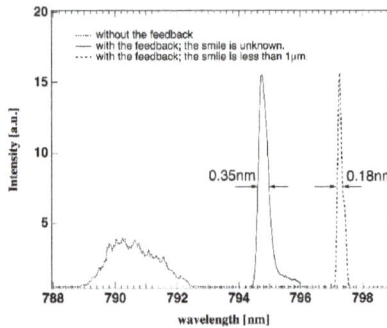

Figure 10. The measured laser spectra with and without the feedback to narrow the spectrum.

Then we assembled a setup for the in situ SEOP ^{3}He NSF with the developed laser optics, and applied it the neutron beam experiments at the polarized neutron reflectometer, SHARAKU (BL17) [26], and the small-angle neutron scattering instrument, TAIKAN (BL15) [27], at MLF, where an ^{3}He polarization degree of 0.68 was achieved and maintained during the experiments [26,27].

2.3.2. A Non-Magnetic Flexible Heater

In SEOP [28], the number density of alkali metal vapor is one of the important parameters to efficiently polarize ^{3}He nuclei. The appropriate number density is on the order of 10^{15}/cm^3 and the corresponding temperature is between 160 °C and 250 °C depending on the K and Rb mixture in alkali-hybrid SEOP [29]. Hot air blowers, which introduce no magnetic fields to the ^{3}He cells, are commonly used to keep ^{3}He cells at appropriate temperatures since ^{3}He nuclei are extremely sensitive to magnetic field gradients and the polarization can easily be lost to unwanted stray fields. Instead of hot air blowers, electrical heaters are occasionally used with extreme caution not to interfere the magnetic field at the cell position [30,31]. Using such an electrical heater allows one to reduce the size of a neutron spin filter as well as enhance safety compared to the use of a hot air blower.

Our non-magnetic flexible heater has two heating layers made of Ni–Cr foil with specially designed strip patterns. The two heating layers are separated and covered by polyimide film for electrical insulation, as illustrated in Figure 11. The two layers of the Ni–Cr foil have an identical strip pattern, and the electrical current through each strip is in the opposite direction so that the magnetic fields produced by the heater currents cancel each other.

Figure 11. The structure of non-magnetic flexible heater. Two layers of Ni–Cr foil are sandwiched between a polyimide film for electrical insulation. The overall thickness of the heater is less than 0.5 mm.

The non-magnetic flexible heater is wrapped on a cylindrical aluminum oven with a diameter of 120 mm and a length of 150 mm to heat a ^{3}He cell inside it (Figure 12).

Figure 12. The non-magnetic flexible heater on a cylindrical aluminum oven. The heater is tightened with non-magnetic brass wires. Additional polyimide foam insulator, which is removed for the picture, covers the heater.

The performance of the non-magnetic flexible heater was measured using a polarized ^{3}He cell placed inside the aluminum oven with the heater current on. Figure 13 shows the spin-relaxation rate measured for a ^{3}He cell to different electrical currents, and it was found that there were no influences on the ^{3}He spin relaxation by the heater current. Note that the operating heater current is less than 1 A for the actual alkali-hybrid SEOP.

We have successfully developed a non-magnetic flexible heater with an excellent performance for the SEOP applications [32]. Such non-magnetic flexible heaters can also be used in other applications that require no magnetic interferences.

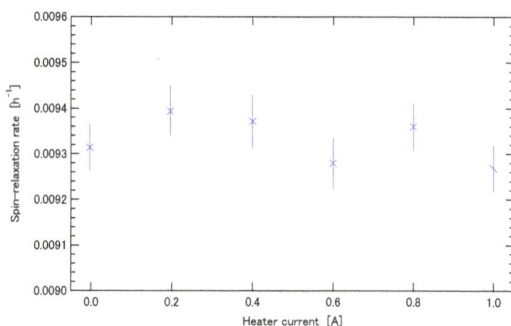

Figure 13. The spin-relaxation time of a ^{3}He cell was measured to the heater current. No current dependence was observed.

2.4. Choppers

2.4.1. T0 Choppers

A T0 chopper is equipped with a massive blade to reduce background noise originating from the high-energy neutron burst. To utilize eV neutrons, we developed a T0 chopper rotating up to 100 Hz [33]. On the high-resolution chopper spectrometer (HRC) at BL12, the dimension of the chopper blade (shielding part) on the rotor is 78 × 78 × 300 mm (300 mm is the length along the neutron beam) with a margin of $\Delta w = \pm 1$ mm (the beam cross section is 76 × 76 mm). This margin corresponds to the phase control accuracy of $\Delta t = \pm 5$ μs at $f = 100$ Hz ($\Delta t = \Delta w / 2\pi R f$), for the rotational radius $R = 300$ mm. Since this T0 chopper is installed 9 m from the neutron source, neutrons with energies up to 2.5 eV can be utilized. In this T0 chopper, the rotor made of Inconel X 750 is mounted inside the vacuum, the shaft is supported by ball bearings, and the rotation is transferred through a magnetic seal from the motor located outside the vacuum chamber, as shown in Figure 14a. The length of the T0 chopper along the neutron beam is 1.3 m. A continuous running time of more than 1000 h and a total running time of more than 4000 h are required without changing components, and we confirmed these requirements. The maintenance should be scheduled considering these periods. The performance of the T0 chopper on HRC is indicated in Figure 14b. A neutron beam monochromatized by a Fermi chopper rotating at 600 Hz is incident upon a vanadium (V) sample and scattering neutrons are measured as a function of TOF. The peaks at around 1940, 4200, and 6460 μs, which correspond to 502, 108, and 45 meV, respectively, are of properly monochromatized neutrons (the peaks at 810 and 3070 μs are from half-turns of the rotation). The background noise is successfully reduced by 2 orders of magnitude at around 500 meV. A variety of requirements from instruments can be covered by a small modification of the structure: a short model less than 1 m long with $f = 50$ Hz for the total scattering instrument NOVA at BL21, and a two-beamhole model with $f = 25$ Hz for the reflectometer SOFIA at BL16. Also, this type of T0 chopper is used in BL04, BL22, and BL23.

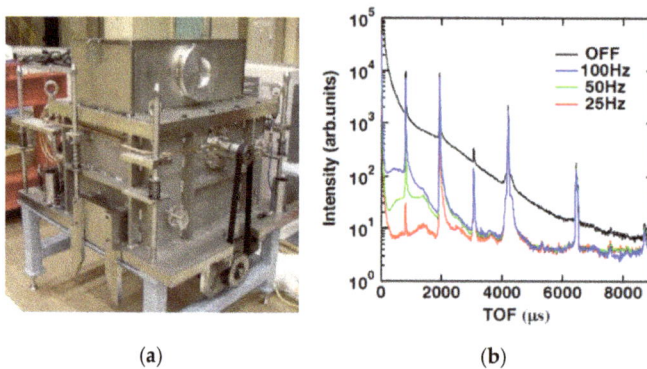

(a)   (b)

Figure 14. (a) Photograph of the T0 chopper installed on high resolution chopper spectrometer HRC. (b) Effect of background noise reduction. Time-of-flight (TOF) spectra for the operation at $f = 100$ Hz, 50 Hz, 25 Hz, and in no operation (OFF) condition are indicated.

Another type of T0 chopper is also being used at MLF. The development target for this T0 chopper was to make the length along the beam path as short as possible, which enables us to minimize the loss of neutron transport. To make our target a success we introduce an in-wheel type motor to the T0 chopper. This motor has an inner stationary portion (the stator), and an outer rotating portion (the outer rotor) that rotates around the stator and drives a chopper blade (Inconel X 750) attached to the outer rotor.

Many years of research and development in collaboration with Kobe Steel Co., Ltd., (Kobe, Japan) (KOBELCO) have accomplished the compact type T0 chopper system [34]. The dimension of the T0 chopper blade for BL01 is 84 mm (cross section) × 300 mm (length along the beam). Each beamline (BL) has a different cross section, determined by the beam size at the location of a T0 chopper. The first product of a compact type T0 chopper is shown in Figure 15a. The length along the beam path is reduced to 500 mm, and the upper limit of rotating frequency is 50 Hz. The inelastic neutron scattering data measured on 4SEASONS at BL01 with and without a compact type T0 chopper are shown in Figure 15b,c, respectively. The background in the high-energy region is significantly suppressed [35]. The compact type T0 chopper is used at BL01, BL09, BL11, BL15, BL18, and BL20. We are now working on research and development of the compact type T0 chopper, aiming at increasing the upper limit of the rotating frequency, which further extends the usable incident energies toward the higher energy side.

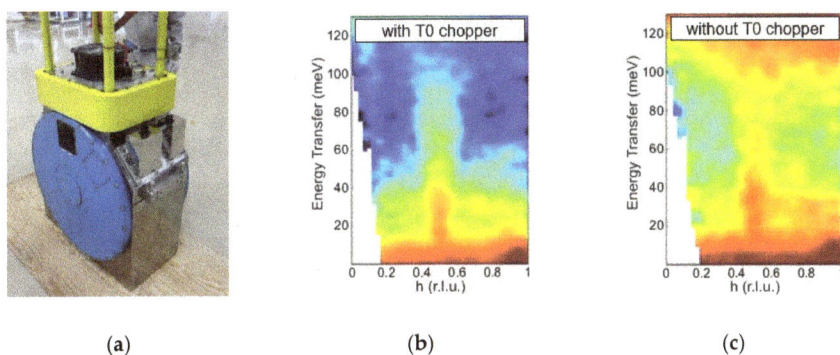

Figure 15. (**a**) Photograph of the compact type T0 chopper for 4SEASONS. Inelastic neutron scattering data obtained by 4SEASONS (**b**) with and (**c**) without the compact type T0 chopper.

2.4.2. Fermi Choppers

A Fermi chopper consists of a rotating slit package to monochromatize neutron beam [36]. We developed a Fermi chopper (Figure 16a) [37] by using a magnetic bearing mounted in a turbo molecular pump whose pumping speed is 1300 L/s, because the mass and the moment of inertia of its rotor blade are almost identical to those of the slit package of the Fermi chopper. This system can be operated in a range of rotational frequencies (f = 100–600 Hz). On HRC, the energy resolution with respect to the incident neutron energy is $\Delta E/E_i$ = 2.5% for the optimum condition $\Delta t_{ch} = \Delta t_m$, where Δt_{ch} is the chopper open time and Δt_m is the pulse width, and, further, $\Delta E/E_i$ = 1% is achieved if Δt_{ch} = 1 µs is realized for eV neutrons. The phase control accuracy Δt should be less than 30% of Δt_{ch} to minimize intensity loss. We assumed that $\Delta t \leq 0.3$ µs, i.e., a frequency resolution $\Delta f = |df/dt| \Delta t = f^2 \Delta t \leq 0.01$ Hz at f = 600 Hz, as the goal of the development. The phase control system was developed using a direct digital synthesizer system with a frequency control unit of 0.00058 Hz at f = 600 Hz. The phase control accuracy in the developed Fermi chopper was observed to be Δt = 0.08 µs at f = 600 Hz, well within the specification of the goal. An inelastic neutron spectrum from a vanadium sample measured by using this Fermi chopper on HRC is shown in Figure 16b, where the energy resolution for E_i = 206 meV is identical to the designed standard resolution of $\Delta E/E_i$ = 2.5%. This type of Fermi chopper is also used on BL21 and BL23.

Furthermore, we have proposed a new method of inelastic neutron scattering measurement in chopper spectrometer by simultaneously utilizing the multiple incident energies (E_i) [38]. This method (Multi-E_i measurement) can reduce the dead time in time-of-flight measurement, and thus can markedly increase the measurement efficiency. Generally, a Fermi chopper is not suitable for Multi-E_i

measurement, because transmission of a Fermi chopper is usually optimized to a specific E_i by curved slits. On 4SEASONS, however, we decided to adopt a Fermi chopper with straight slits that sacrifices the transmitted intensity but passes through a wide energy range of E_i. This decision brought us success in the experimental demonstration of the usefulness of Multi-E_i measurement, even in the Fermi chopper spectrometer [39].

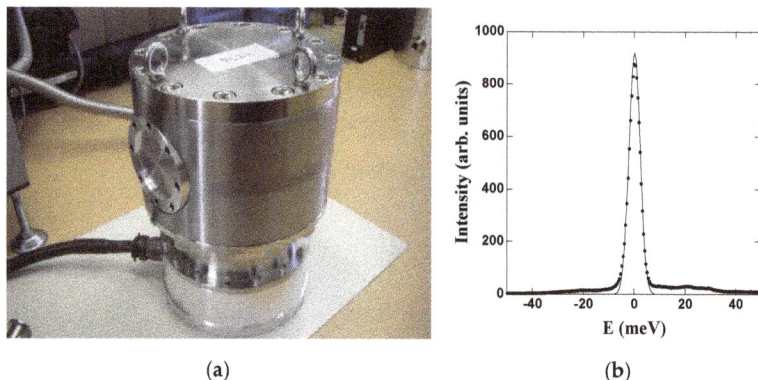

(a)   (b)

Figure 16. (a) Photograph of Fermi chopper. (b) Observed inelastic scattering spectrum from a vanadium sample for $E_i = 206$ meV with the Fermi chopper rotating at $f = 600$ Hz on HRC. The FWHM of the peak is $\Delta E = 5.4$ meV.

We have continuously re-examined the performance of the Fermi chopper. Recently, we replaced the existing Fermi chopper of 4SEASONS with a newly developed one that has a more compact slit package [40]. The new Fermi chopper provides us with more neutron flux without lowering the resolution. The inelastic neutron scattering data of vitreous silica, measured at 4SEASONS using the previous Fermi chopper and the new one, are compared in Figure 17. The measurement time was 30 min for both.

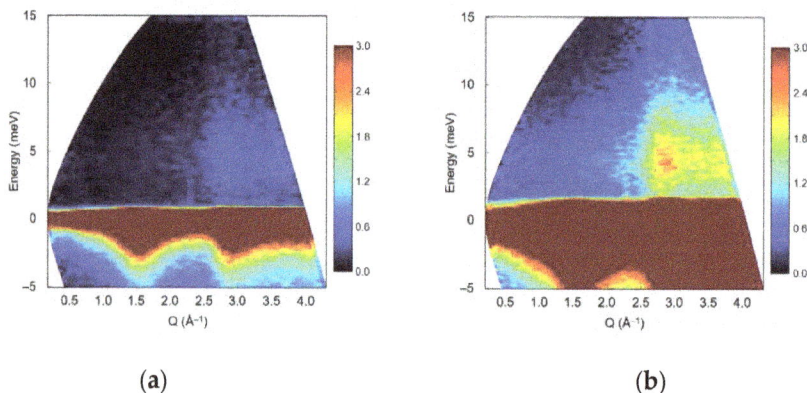

(a)   (b)

Figure 17. *Cont.*

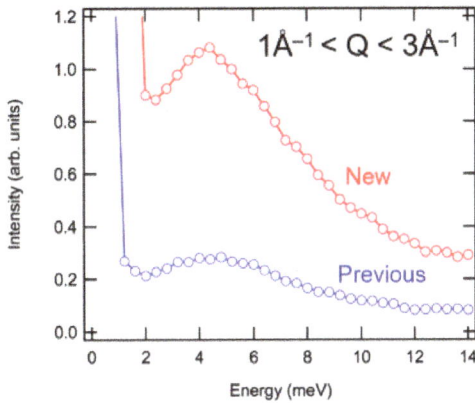

Figure 17. Inelastic neutron scattering spectra of vitreous silica measured at 4SEASONS by using (a) the previous Fermi chopper and (b) the new one. The measurement time was 30 min for both. (c) A comparison of the energy dependences of $S(Q, E)$ between the previous Fermi chopper and the new one.

2.4.3. Disk Choppers

Neutron disk-choppers are simple but indispensable devices for neutron spectrometers. Therefore, most of the neutron instruments at MLF employ disk-choppers. The disk-choppers employed at constructed and planned neutron instruments at MLF are listed in Table 1. At MLF, two types of disk-choppers are used. One is a slow disk-chopper, which runs at a relatively slow speed, $f \leq 50$ Hz, and is used to define the band width or frame and eliminate undesired neutrons. The other one is a fast disk-chopper, the maximum revolution of which can exceed 300 Hz, and is used in a monochromator or for shaping the pulse width of the neutron beam coming from source. Pulse shaping is a rather special task of the fast disk-choppers at MLF, since some inelastic instruments utilize high intensity from the coupled moderator source by shaping a broad pulse shape from this type of source by fast disk-choppers to realize high intensity and high resolution simultaneously. Therefore, as much effort has been devoted to fast disk-choppers as to slow disk-choppers.

Two main benders, KOBELCO and MEISYO KIKO, provide disk-choppers to the instruments at MLF. There we try to unify the menu terms, look, and feel so as not to confuse the two.

Commissioning of disk-choppers was conducted by an MLF chopper task team. The electronic noise level, which can affect other devices in the experimental halls, and performance (phase stability, transmission of the disks, temperature evolution of spindles and housing, etc.) were investigated for all disk-choppers installed in instruments at MLF. The information is shared by all instrumental teams through this task team.

Slow Disk-Choppers

The slow-speed disk chopper was developed for the device as a band definition chopper or a frame-overlap suppression chopper or a pulse suppression chopper, for example at 12.5 Hz.

The prototype slow disk chopper was developed from 2004 with KOBELCO. Its main specifications are as follows.

(i)　For the ease of maintenance, the neutron beam is passed under part of the disk chopper, and the body of chopper disk and motor are mounted on the stand (see Figure 18).

(ii)   For the reduction of disk weight, the disk was made of an Al alloy applied with (isotope enriched $^{10}B_4C$ powder + epoxy resin) as a neutron absorber.

(iii)   The transmission of neutrons was less than 10^{-6} at E_i = 100 meV.

Nowadays many slow disk choppers are produced and installed in many beamlines, which are designed referring to the prototype slow disk chopper, even in those produced by MEISYO (see Table 1). On the slow single disk chopper installed in BL19, we performed some commissioning tests. Figure 19 shows the commissioning results. From this phase control test, we determined the operational parameter for this disk chopper.

Table 1. Disk Choppers at the Materials Life Science Facility (MLF).

Beam Line	Type of Chopper (No.)	Manufacturers
BL01	SI(2)	KOBELCO
BL02	FII(1), FI(2), SII(2), SI(1)	KOBELCO
BL03	SI(1)	MEISYO
BL04	SII(1)	MEISYO
BL05	none	-
BL06	SI(2)	Vacuum Products.
BL08	SII(1), SI(2)	MEISYO
BL09	SI(3)	MEISYO
BL10	SI(1)	MEISYO
BL11	SI(2)	KOBELCO
BL12	none	-
BL14	FII(2), FI(2), SII(2)	KOBELCO
BL15	SI(3)	KOBELCO
BL16	SI(1)	Vacuum Products.
BL17	SI(3)	KOBELCO
BL18	SI(2)	KOBELCO
BL19	SI(1)	KOBELCO
BL20	SII(1), SI(2)	MEISYO
BL21	SI(1)	MEISYO
BL22	SII(1)	MEISYO
BL23	FI(1) correlation, SI(2)	MEISYO

S: slow chopper; F: fast chopper; I: single disk; II: double disk (number of that type chopper in the beamline).

Figure 18. The slow single disk chopper installed in BL19.

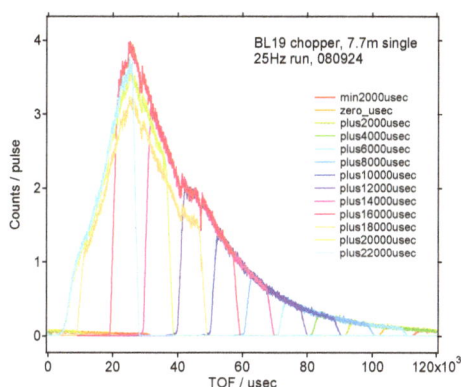

Figure 19. Phase control test on the slow single chopper in BL19. It shows the incident beam TOF spectrum selected by the disk chopper in several phase delay conditions.

Fast Disk-Choppers

The development of fast disk-choppers started in 2004 with KOBELCO. The cold-neutron disk-chopper spectrometer AMATERAS (BL14) requires a minimum burst time of less than 8 μs to realize 1% energy resolution. The requirement was that a disk of 600 mm diameter with a 1 cm slit should run at 350 Hz revolution. In 2004, such disk-choppers were not available, which is why we started developing them. There were several technical issues during the development. The most important key is the disk itself. After experiencing several crashes of disks and sometimes even a full chopper system, finally disks can be run at a revolution of >350Hz (Figure 20a). Special care was taken in optimizing the direction of carbon fibers of carbon fiber reinforced plastic (CFRP) and the method of placing. Also, we used the metal ^{10}B, not ^{10}B$_4$C, to reduce the mass at the edge of the disk.

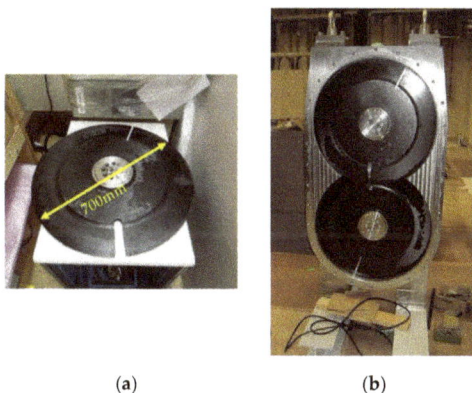

(a)　　　　　　　(b)

Figure 20. (a) A disk for fast-disk choppers and (b) inside view of No. 2 fast-disk chopper of AMATERAS.

At MLF, three sets of fast disk-choppers have been installed at AMATERAS. Specifications of the three choppers of AMATERAS are listed in Table 2. No. 1 chopper shapes the width of a pulse from the source. No. 2 chopper is the monochromator chopper. No. 3 chopper is used for removing unwanted neutrons coming through No. 1 choppers. Motors and disks of No. 2 chopper are placed one above the other to set two disks as close together as possible (Figure 20b).

This is essential to perform multi-E_i measurements on AMATERAS [38]. Commissioning of fast disk-choppers has been done in the course of the commissioning of AMATERAS. By using these choppers, we have confirmed $\Delta E/E_i \sim 1\%$ energy resolution in the range of $E_i \leq 3$ meV.

Based on the experience of developing fast disk-choppers of AMATERAS, in 2011 we manufactured other fast disk-choppers for the biomolecular dynamics spectrometer DNA (BL02). DNA uses three sets of fast disk-choppers (one of 225 Hz—double-disk type with 4 slits (1 cm × 1 slit and 3 cm × 3 slits) for pulse shaping and two of 150 Hz—single-disk type with 4 and 2 wide slits, respectively, for removing unwanted neutrons); these choppers have been working since 2012.

Table 2. Fast disk-choppers installed at AMATERAS and DNA.

AMATERAS	No. 1	No. 2	No. 3
$L_{\text{modrator-chopper}}$	7.1 m	28.4 m	14.2 m
Disk Radius	350 mm	350 mm	350 mm
Revolution	≤350 Hz	≤350 Hz	≤350 Hz
No. of Disks	2 (Counter-Rotating)	2 (Counter-Rotating)	1
Slit Width	30 mm	10 & 30 mm	30 mm
Min. Burst Time	22.7 µs	7.6 µs	45.5 µs
Gap between Disks	50 mm	20 mm	-
DNA	**No. 1**	**No. 2**	**No. 3**
$L_{\text{modrator-chopper}}$	7.750 m	11.625 m	23.250 m
Disk Radius	350 mm	350 mm	350 mm
Revolution	≤225 Hz	≤150 Hz	≤150 Hz
No. of Disks	2 (Counter-Rotating)	1	1
Slit Width (deg.)	1.9 (10mm) & 5.7 (30 mm)	24.7	56.6
No. of Slits Min. Burst	4	4	2
Time	12 µs & 36 µs	-	-
Gap between Disks	50 mm	-	-

3. Computational Environment

Software now plays an indispensable role in scientific research in all fields. In neutron and muon measurement, software affects the efficiencies of all processes such as planning of measurements, optimization of hardware condition, hardware control, collection of data, data correction of hardware dependent factors, model constructions, comparison with theories, and so on.

The MLF computational environment was started in 2003. Five software components were defined: experiment, analysis, simulation, database components, and a user interface to integrate these components. Under a collaboration with KEK Computing Research Center (Tsukuba, Japan) a prototype of the data analysis framework with C++ was built in 2004 as the core of the analysis component [41]. It was four years before the operation of MLF. Until day 1 of MLF in 2008, two components were developed: (1) core of the experiment component, DAQ-Middleware for MLF neutron scattering [42], in the collaboration with the Particle and Nuclear Laboratory, KEK; and (2) a user interface component (IROHA). The network infrastructure of MLF was established by the J-PARC information section. Since then, each component has been developed.

International collaboration has been pursued with ISIS facility, Rutherford Appleton Laboratory and Spallation Neutron Source (SNS), Oak Ridge National Laboratory (Oak Ridge, TN, USA) [43]. Unfortunately, the collaboration is not successful so far since the software development was insufficient at MLF then. However, to catch up with the very rapid progress of software, international collaboration is still desired.

In this section, the described MLF computational environment focused on the neutron scattering facility. The computational environment of the Muon facility is different but some of the components, such as the user interface (IROHA2, described in Section IROHA2), will be used as common components.

3.1. Main Components of MLF Computational Environment

3.1.1. Instrument Control Software Framework: IROHA

Overview of Instrument Control Software

The software for instrument control should be user-friendly, scalable, and flexible for MLF experiments. To satisfy these requirements, we have developed a common software framework, called IROHA, for instrument control at MLF [44,45]. The characteristics of IROHA are as follows:

- Scalability: high throughput for large-scale data on the gigabyte order.
- Flexibility: adaptability of various experimental purposes by providing hardware control software for a variety of hardware.
- User-friendly: graphical user interface and controllable via remote access.
- Automation: programing measurement with graphical user interface as well as command-line interface.

IROHA

During the construction period of MLF, we have developed an instrument control software, IROHA, which consists of the common graphical user interface (GUI) software called Working Desktop (WD) and experimental controlling software to operate DAQ hardware and other equipment such as SE equipment and beamline optics components. The source code of IROHA was written in an object-orientated script language, Python, and XML/HTTP was chosen as the protocol between software components. Users can perform their experiments and data analysis by sending Python commands through IROHA. In other words, users can seamlessly operate experimental controlling, data analysis, and visualization with one Python script. IROHA is introduced to the many instruments of MLF and currently used in MLF users' experiments.

IROHA2

Several years ago, we started to upgrade IROHA to improve the following points [46,47]:

1. Proper role-sharing between the software of individual device control and that of the integrated instrument control.
2. The interface with database systems such as MLF experimental database (MLF EXP-DB), the authentication system, the user information database cooperated with the system of J-PARC Users' office for linking metadata with experimental data.
3. Platform-independent user interface.

The new software framework for instrument control at MLF, named IROHA2, was first released in 2013 and introduced to several instruments and users' experiments.

For proper role-sharing of software, IROHA2 consists of four software components:

- Device control server: operation, monitoring and logging outputs of devices. More than 50 devices have been supported.
- Instrument management server: generation of run information, configuration and management of instrument components and user authentication.
- Sequence management server: configuration and management of automatic measurement.
- Integrate control server: integral operation and monitoring of an instrument; monitoring is also available from outside of MLF.

The interfaces with database systems in IROHA2 are the main function of the instrument management server. The instrument management server generates run information included the URI (Uniform Resource Identifier) of the measurement data (from the DAQ system), the user ID, the title of the experiment, the sample information (from the user information database), device information (from the device control servers), and the measurement time of each measurement. Because run information is written in XML format, we can describe various data flexibly. MLF EXP-DB collects and catalogs the run information for the management of experimental data.

We have introduced a web interface as a platform-independent user interface for IROHA2. Users can access IROHA2 not only from desktop computers but also from mobile devices such as tablets and smartphones with a web browser on each operating system. More than 50 devices have been supported by the device control server of IROHA2.

Issues with IROHA2 are follows: (1) supporting more devices, especially users' devices brought to MLF; (2) introduction of a real-time communication method such as a message queue, instead of recording data files, for 1 MW operation; (3) continuous improvements towards a user-friendly interface.

3.1.2. Data Acquisition

Event Recording Method

The standard DAQ mode at MLF is the event recording method [48,49]. The event recording method has realized that each signal that is a detection of neutrons and muons, conditions of SE equipment (position of stage, angle of goniometer, temperature, pressure, magnetic and electric field, etc.), status of beamline optical components (injection beam intensity, slit size, phase delay of chopper, neutron polarization, etc.), timing of proton injections, and so on, is recorded as an event with the timestamp. We have assembled the event recording system for these various events as shown in Figure 21. Many kinds of signals are input to the DAQ boards (GateNET, TrigNET, NeuNET, and Readout) and converted into event data. The event data are collected and stored by the DAQ-Middleware running in DAQ computers.

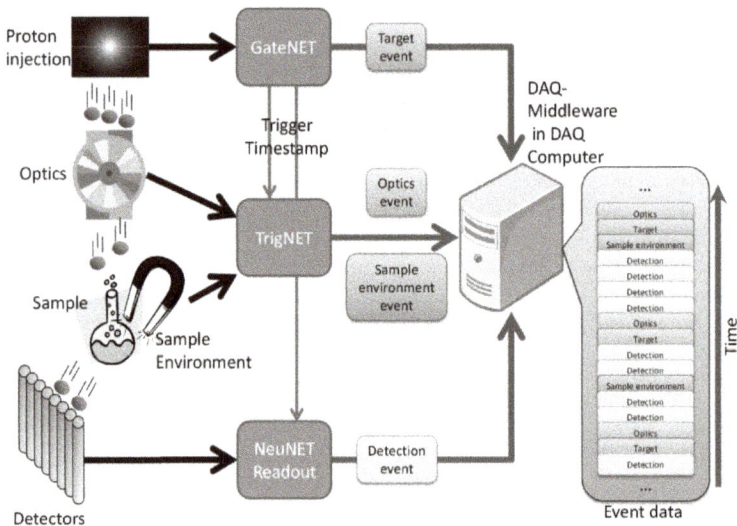

Figure 21. Diagram of event recording method.

High-throughput performance is necessary in the event recording method in a high-intensity facility like MLF. We have realized this by introducing a distributed network environment into the DAQ boards and software. The boards have incorporated a SiTCP Ethernet interface [50], which is fieldbus network protocol technology. The DAQ software incorporated DAQ-Middleware, which is easily configurable DAQ software under a distributed network environment.

The event recording method realized data reduction of neutron scattering combined with every recorded events. This is described in the data reduction section.

Data Acquisition Software

DAQ-Middleware [51–53] is a data acquisition software framework developed at KEK. We can build the DAQ software by assembling several DAQ components of DAQ-Middleware in flexible and scalable. In addition, DAQ-Middleware can make the DAQ component include customized logic. Because these characteristics satisfy our demands, we have adopted DAQ-Middleware as the standard DAQ software at MLF, installed on all the DAQ computers of the many MLF instruments. We have developed various readout components according to detectors equipped in MLF instruments and the components to produce data for monitoring and analysis.

3.1.3. Data Reduction and Visualization

Overview

In general, there are many procedures for data analysis to obtain scientific results from measured data. The MLF computational environment group clearly separates these procedures into two parts, as below.

1. Data Reduction

    -   Conversion from the binary raw data, i.e., the event recording data at MLF, to the histograms format.
    -   Data correction, for example the detector efficiency, the solid angle, the background subtraction, the calculation of absolute value on the structure factor, and so on.

2. Data Analysis

    -   Interpretation of measured data, like structure refinement and simulation calculation.

Our group mainly produced the data reduction software by ourselves because we regard it as a part of hardware of instruments. It is expected that a deep commitment to data reduction will make it possible to quickly catch up with advanced hardware technologies and to develop novel measurement and analysis methods for MLF users. For data analysis, on the other hand, we decided that users should utilize not only our software but also external popular software because the analysis software is strongly dependent on a user's particular type of science and we have poor development resources. We, therefore, offer some conversion software from our data format to that of external software. In addition, we also produced the basic visualization software ourselves, while the advanced visualizations are treated as a part of the data analysis.

We developed framework software for data reduction and analysis to unify data formats and functions at the MLF facility at first. Next, we developed some of the data reduction and visualization software on this framework. These results are mentioned in the following sections.

Object-Oriented Data Reduction and Analysis Framework Manyo Library

Many neutron scattering instruments have been installed at MLF, but requirements of data reduction and analysis software for each instrument are quite different among instruments. The MLF computational group has developed a software framework for developing data reduction and analysis software that can be applied to each type of instrument [41,54,55]. Now the framework Manyo Library is working on the 16 instruments that are shown in Figure 22. The framework provides fundamental and generic functionalities for developing application software for neutron scattering experiments. Because three-dimensional histograms should be handled in the framework, we developed a series of data containers that can be applied to one-, two-, and three-dimensional histograms. ElementContainer (EC) is a simple and fundamental data container in the framework, which can store a one-dimensional histogram with its error values, where any number of vector objects can be kept with their names and metadata. Users can access each data object with its name. Data containers for two- and three-dimensional histograms, ElementContainerArray (ECA) and ElementContainerMatrix (ECM), are also prepared (Figure 23). One-, two-, and three-dimensional histogram data are stored in the EC family with their metadata, and four arithmetic operations between them can be performed with error propagations.

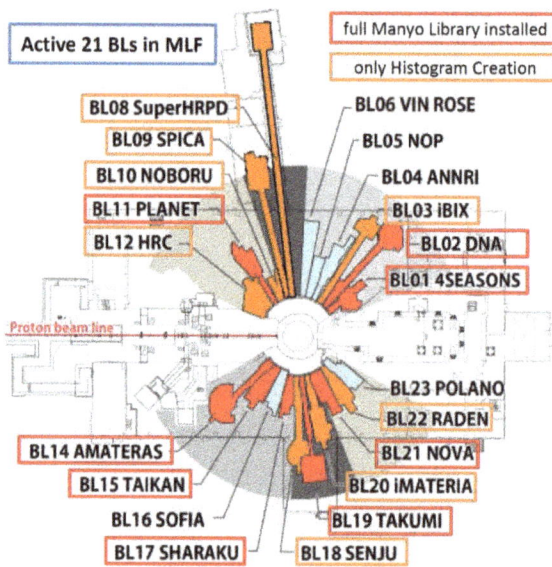

Figure 22. Floor layout of MLF. The Manyo Library is working on the 16 beamlines indicated by orange and red boxes.

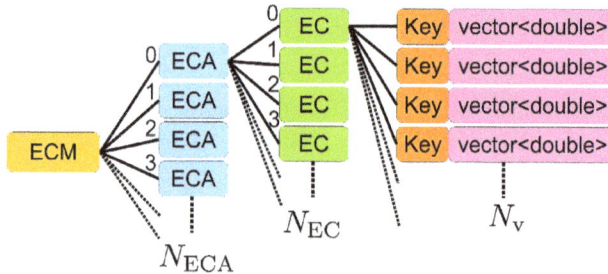

Figure 23. Structure of the ElementContainer (EC) family defined in the framework. The number of vectors, EC and ElementContainerArray (ECA) in an EC, ECA and ECM are N_v, N_{EC} and N_{ECA}, respectively.

Standard File Format: NeXus

We decided that the standard data file format at MLF should be NeXus, and an interface between data containers and NeXus file is prepared in the framework. A simple data file structure in NeXus format was designed in 2016. An interface between the EC family and NeXus files was developed by using the HDF5 C-API library. The rates of reading and writing ECMs from/to NeXus files with the interfaces were obtained and shown in Table 3. The file sizes on the disk are 4.0 and 1.3 gigabytes at (N_v, N_{EC}, N_{ECA}) = (3,100,300) and (3,10000,1), respectively. The data structures and dimensions of the former and latter datasets are the same as those of the three-dimensional histogram data produced by the chopper spectrometer and the powder diffractometer working at MLF. The efficiency of the interface is good enough but we should concern ourselves with how to increase the efficiency of data input/output in the framework, because the size of data files handled by the framework is increasing as the intensity of proton beam producing pulsed neutron beams increases. In future, we will improve the interface with multi-thread or network-distributed-parallel processing. Histogram data with their metadata in the NeXus format can be shared with other facilities around the world using NeXus-API or the HDF5 library.

Table 3. The efficiency of reading and writing NeXus files from/to the ElementContainerMatrices (ECMs). T_w and T_r are the writing and reading times in seconds with the interface. N_v, N_{EC} and N_{ECA} indicate the dimensions of ECM.

N_v	N_{EC}	N_{ECA}	T_w	T_r
3	100	300	24.7	8.33
3	10,000	1	10.39	2.76

Utsusemi, the Base Software for Data Reduction and Visualization at MLF

Overview of Utsusemi: Utsusemi is the software series used to analyze and visualize the neutron scattering data observed at MLF [56]. MLF adopted an event recording method for the DAQ system that records events including the time of flight and the position of the neutron detection. This method gives MLF users higher flexibility and efficiency for measurement and analysis. In event recording, the role of software is very important. The development of Utsusemi started from the software development for the chopper-type neutron inelastic measurement, which includes the device control system, data analysis, and visualization. These development products were utilized in actual user measurements at the chopper spectrometers at MLF, while we were proceeding with development of effective analysis for event recording data. In particular, we successfully contributed to the realization of the Multi-E_i method for inelastic scattering for the first time in the world [39]. As the other beamlines also required event-recording analysis, we made the data reduction and visualization parts in Utsusemi independent of the original to give it greater transferability as universal analysis software at MLF. We also produced additional packages based on Ustsusemi for each beamline. As a result, Utsusemi is introduced in a lot of beamlines at MLF to become one of the frameworks for data reduction and analysis.

Utsusemi software is written in C++, based on the Manyo Library framework and Python. All main functions are included in C++ codes for the processing speed and can be used from the Python interface. Utsusemi also has the GUI written in Python code and uses Python libraries. Because the user interface of Utsusemi is unified in Python, it is easy for users to make new functions or analysis procedures tailored to users' specific needs with Utsusemi commands.

Utsusemi software has already been distributed to many users at MLF with a release number. However, we think the current Utsusemi has a lot of functions that need to be improved or developed, so we continue to deliver the updated one with a short span, the so-called rolling release concept. For this release concept, we have a plan to build a portal site for the MLF computational environment group and a download page to distribute our updated software, including Utsusemi.

Data Reduction for Event-Recording: The most important function of Utsusemi is conversion from event-recording data produced from the MLF DAQ system to histogram data for easy understanding and visualization of the measurement results. This conversion function is very flexible. Using the information of event data like TOF, position of neutron detections, and absolute time, users can make histogram data from any TOF range and time region from measurement data. This enables users to easily measure the time-transient phenomena of the sample. For example, this function is used for a data reduction of an in situ neutron diffraction measurement during time-dependent temperature and tensile stress control, so called thermo-mechanical simulation (see Figure 8 of [57]). In addition, we successfully developed a DAQ electric module, TrigNET, which records common electric signals as events. By analyzing TrigNET events connected with instruments and SE equipment, neutron events can be filtered for each condition. We have already achieved this advanced method in the actual experiment and analysis on Utsusemi, and we provided this for users—for example, the measurement of the sample under a pulsed magnetic field and periodical electric field. Another example is a stroboscopic data reduction of the measurement of piezoelectric material under a cyclic electric field [58]. The Utsusemi event conversion supports most types of detectors utilized at MLF—for example, ^{3}He gas position sensitive detector, 1D- and 2D scintillation counters, and so on.

Recently, we succeeded in adding a differential event data reduction function to enable quasi-real-time analysis. During the measurement, the DAQ system continuously stores measured event data as binary files on a hard disk. This function only reads the difference between current binary files and previously read ones, with an interval time to be converted to histograms. Using the differential data, users can see the real-time change in measured data using visualization. Because this differential data format is the same as that of static event data, the analysis codes can also be unified between the two methods. This method has already been introduced in many beamlines to help users.

Command Sequence Execution: This software, named Sequence Editor, can be accessed via a GUI (Figure 24) and helps users to make procedures for executing analysis commands. Users can make their own command sequences to execute data reduction and visualization in this software. Commands are prepared by the instrument staff. Arguments for selected commands can be changed freely by editing the text boxes at the bottom of the window. These argument boxes are dynamically constructed according to a selected command. By pushing the Start button, the Sequence Editor executes a command sequence, step by step. If any error occurs in executing a command or the process is interrupted by the Stop button, the process stops but this software keeps all the parameters. All resulting data from the commands are kept in memory until they are deleted and users can directly send data to visualization software from the Sequence Editor. In addition, the command sequence made by users can be saved as a Python script to be executed directly on the console. If required, users can run the script directly in the Utsusemi environment or improve the saved script for their own analysis.

Visualization: Visualization software helps users to see and check any data at each data reduction step, from a simple TOF data to multi-dimensional data, i.e., $S(Q, E)$, measured by inelastic neutron scattering instruments. Utsusemi visualization software mainly consists of 1D-plotter (X and Y axes), 2D-plotter (X, Y axes and the intensity), PSD map visualization, and the data slicing software for multi-dimensional space data for a single crystal sample. These plotters can be controlled through the GUI or command-line on Python.

1D-Plotter, named MPlot, is usually used just to plot several histograms stored in data containers. This plotter has several basic functions such as changing types of lines and markers, scaling plot region, import/export of any text data, and so on. 2D-Plotter, named M2Plot, can plot two-dimensional data like the powder/grassy samples $S(Q, E)$ data and the sliced multi-dimensional data of a single crystal sample. VisualContM software treats multi-dimensional data $S(Q, E)$ obtained by a single crystal measurement This software makes $S(Q, E)$ data from the $S(phi, E)$ data produced from each detector histogram using sample information, i.e., the lattice parameters and the orientation against the incident beam. VisualContM has data slice functions to send the sliced data to M2Plot to visualize. We also produced another kind of multi-dimensional slicing software that combines several tens of datasets measured at different orientations of a single crystal sample to cover a wider area in Q-space, which is useful for a 3D spin system. This software, named D4matSlicer, merges a series of datasets into one $S(Q, E)$ data matrix and enables us to slice it.

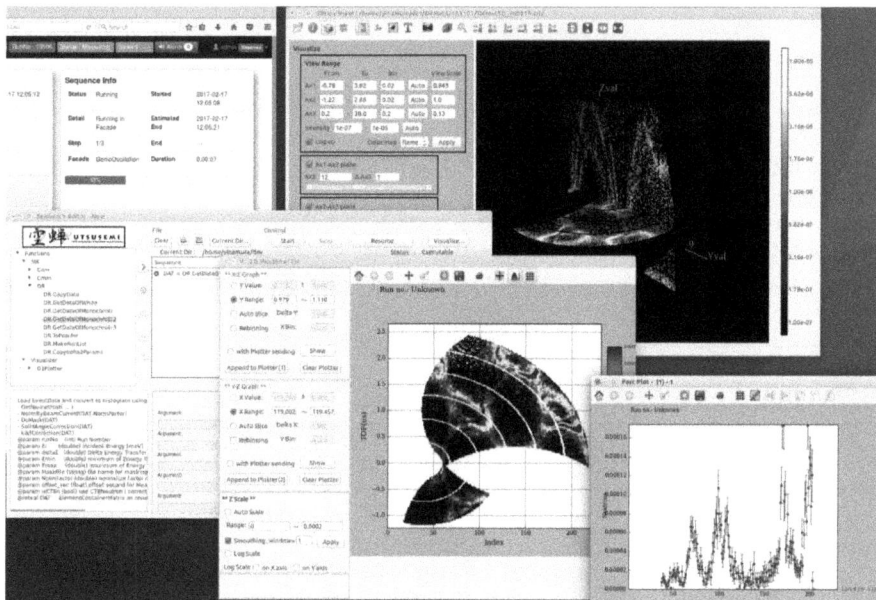

Figure 24. A screenshot of Utsusemi running.

Usability Improvement: General users have had difficulty installing and using Utsuemi on their own PCs until recently because the software and codes run only on Linux OS. This means that users must construct a Linux environment before using our software. Such a complex installation hinders the users' ability to analyze data at their offices and quickly prepare their results. To solve this problem, we produced a binary application and installer software executable on Windows and Mac OS. The binary installers are easy to use because they follow a regular way to install an application into each operating system (OS) and include all binary applications of Utsusemi, including Manyo Library, with the required software libraries. They work on 64-bit Windows 7, 8.2, and 10, or Mac OS 10.10 (Yosemite) and 10.11 (El Capitan).

Remote Access Framework, Nokiba and Shigure: At MLF, a remote access system is under development and partly in the test phase. The remote access system that we are developing aims to make it possible for remote users to process data. For system flexibility, our remote access system consists of two separated systems. One is a web interface that is used through the web browser and the other is a processing backend. The web interface is named Nokiba. Nokiba is a server-side program written in Python and run with a web server. The processing backend is Shigure. In typical usage, users can send commands to Shigure through the Nokiba web interface. Nokiba and Shigure are tied with a network connection and communicate with AMQP. AMQP is an abbreviated form of Advanced Message Queuing Protocol, which is used for message queuing and routing.

Shigure works under the publish-and-subscribe (pub/sub) model with AMQP. The pub/sub model is an asynchronous messaging paradigm. In that model, a sender of message does not know the implementation of receiver(s). In the pub/sub model of Nokiba and Shigure, users make commands for processing and send the commands to the Shigure backend through the Nokiba web interface.

The server side of Nokiba sends the message to the message broker over AMQP. In this scheme, the Nokiba server acts as a client (publisher) in the Shigure pub/sub model. Subscribers to Shigure can process Python script. One benefit of the pub/sub model is its flexibility. This loose coupling of Nokiba and Shigure enables us to adjust processes depending on the number of users or processes. This means it can start at a small scale and later increase in stages.

In the first stage of Nokiba and Shigure, data reduction with Utsusemi can be done through a web browser. Now we have developed a web user interface in prototype and Utsusemi can run on the Shigure backend. We started on improvement of the stability of the pub/sub implementation. On the other hand, this type of remote processing system has already been implemented in commercial cloud services (e.g., Amazon Web Services, Google Cloud Platform, etc.) and the utilization of such a commercial cloud will be beneficial for the reduction of costs and manpower. We are implementing the system in the cloud services for reviewing.

Issues with the Data Reduction Software: As described above, our data reduction software has been steadily developed to reach a high quality. However, many problems remain to be solved. One is the lack of advanced data correction functions—for example, the calculation of instrument resolution for obtained data, the correction for the neutron absorption process, the multi-scattering process, and the multi-phonon process at the sample. Another is the insufficiency of the software optimized to the target science for each beamline. Since the Manyo Library and Utsusemi provide the framework for the data reduction and include many common functions, we must prepare software suitable to users and their target science. Though users require these problems to be resolved, we cannot assign enough persons to these developments. For the sake of human resources, we have considered making collaborator members responsible for the beamline or allowing expert users to support software development. On the other hand, since some beamlines have produced original software for their own purposes, the effective management of such software is also an important issue.

In addition, as the intensity of MLF neutron source increases, the data size obtained in users' experiments becomes larger. We are afraid, therefore, that it will take too long to handle so much data. This technical problem takes a long time to solve. Our group must decide our priorities on the issues of software development and show our developments plan and schedules over the next few years.

3.1.4. Experimental Database: MLF EXP-DB

Overview of Experimental Database

At MLF, a large amount of experimental data is being produced in rapid succession with high-intensity beams. The typical data size produced in one experiment ranges from several tens of MB to several hundred GB in accordance with instruments and experiments, and the total amount of data generated annually in a whole facility at full performance is on the order of PB. Therefore, data management is an urgent issue for the facility and instrument staff. It is required to safely and efficiently store these data with limited storage resources over a long duration, and furthermore promote the acquisition of scientific results using them. Therefore, the access to and download of measurement logs (experimental metadata) and experimental data is also quite important for facility users. In order to meet these requirements, data should be centrally managed in a unified data repository in the facility, and rapid and effective data access should be provided to facility users. For those reasons, we have developed an integrated data management system, called MLF Experimental database (MLF EXP-DB), as a core system of our data management infrastructure [59]. The data policy of MLF is, that the experimental data will be kept by the facility and open to users for a certain time frame (three years), but the system for opening the data has not been implemented yet.

MLF EXP-DB

The system aims to deliver advanced services for data management and data access at MLF. To develop the system, we have employed Java-based commercial middleware with an XML database system provided by Quatre-i Science Ltd., (Kyoto, Japan), named R&D Chain Management System Software. This middleware is based on a three-layer architecture model and consists of three main components: RCM-Web, RCM-Controller, and RCM-DB. These components all adopt XML technology, so that the middleware has flexibility and extensibility for data structure, workflow, and display output. Since data can have various structures at MLF according to instruments and experimental conditions using various SE equipment (experimental metadata), flexibility is necessary for handling such unstructured and semi-structured data. Moreover, extensibility makes it possible to meet future changes in system requirements. The experimental metadata will be collected through IROHA/IROHA2 and related business database such as proposal, sample, and so on. The main features of the system are as follows:

- Data cataloging: automatically correcting experimental data from instruments and creates the data catalog, which includes information on the paths of data files, samples, measurement conditions, and experimental proposals.
- Central data management: managing data file reposition. It processes the data transfer from instrument local storage to the unified data repository in the facility, archiving for long-term storage and on-demand retrieve.
- Database Link: correcting associated information on experiment such as experimental proposal, primary investigator, and chemical safety of samples by the database link to other business database systems.
- Web portal: providing the interface for data management and data access to facility staff and users via web portals. It is possible to manage, search, and download experimental data in these portals.

Reliability

Recently, we have improved the system, enhancing reliability toward the full-scale operation of the system and facility in the future. In the full-scale operation, the system is required to meet a dramatic rise in data rate with beam intensity and instruments performance resulting in an increasing load of data collection. Additionally, the database becomes rapidly enlarged owing to structural information such as tags and attributes in XML, so that its analytical process becomes lengthier with the increasing number of users. In such a situation, the system is required to provide effective and rapid data access to a large number of facility users, which is estimated to reach 10,000 annually. Therefore, the following three requirements should be satisfied [60]:

1. High reliability is required for a core system; however, the conventional system runs on a single physical server. By removing the single point of failure, service outages should be avoided as much as possible.
2. Scalability for data collection is required to accept the load for data collection increasing with instrument performance and beam intensity. However, the conventional system, which is a single integrated server, does not have such scalability.
3. A web portal enabling effective and quick data access should be provided to facility users. A data search function is especially important to find data for analysis from a large amount of data.

To address these issues, we redesigned the system and improved the web portal as follows:

- High availability: we improved the system as a redundant distributed system in a switch-over relationship. It is possible to perform a continuous data collection and provide stable data access in this configuration against system failure and service outage.
- Scalability: we redesigned the system to a scaled-out configuration enabling data collection load balancing and partitioning of bloating database. The improved system comprises two physical nodes. This architecture enables us to scale performance by adding nodes responding to the data rate.

- Usability: we improved the web portal for data access by implementing a flexible data search function. Users can search data with various conditions such as experimental proposal, sample, device conditions, etc.

Status of Operation

Currently, data management using the system is being performed on a trial basis with some instruments, i.e., BL02, BL11, BL17, and BL18 at MLF. In this trial operation, the data produced by previous experiments are primarily collected to accumulate metadata and verify the required information in the data catalog. We plan to start a full-scale operation soon. For data access, the basic development of the web portal has been completed. We will begin a trial of the web portal and remote access from outside the facility.

4. Sample Environment

The high penetrability of neutrons into materials enables us to prepare various kinds of sample environments (SE) to apply specific experimental conditions. Thus, sample environments are very important, especially for the neutron scattering technique. Typical sample environments used in neutron scattering experiments range from low temperature (up to several mK) to high temperature (up to 1873 K), steady magnetic environments to 10 T and their combination. These sample environments are widely used both in neutron diffraction experiments to determine the structures (atomic arrangements) of materials and in inelastic neutron scattering experiments to study elementary excitations, etc. of materials. On the other hand, there is another type of sample environment that has a dedicated purpose and usage. For example, a load frame to apply stress into a specimen to perform in situ neutron diffraction experiment under stress for studying mechanical properties is such a sample environment.

The strategy for introducing and operating in various sample environments at MLF is as follows: an official SE team is organized for commonly used SE, while an individual neutron instrument group develops and installs dedicated SE. The sample environment team provides technical support to individual neutron instrument groups.

In this chapter, we describe the status of the sample environments at MLF.

4.1. SE Equipment Available at MLF

The SE team is organized to operate and perform maintenance and development on the common SE equipment at MLF. The team consists of JAEA, KEK, and Comprehensive Research Organization for Science and Society (CROSS) staff, who all work together [61,62]. Two kinds of SE equipment are defined at MLF. One is the common SE equipment belonging to the SE team; the other is dedicated SE equipment belonging to individual beamline groups. High-cost SE equipment, high-performance SE equipment and standard SE equipment are the common types. Examples are a high magnetic field system, a dilution cryostat, and a standard cryostat. On the other hand, standard SE equipment such as a 10 K cryostat and special equipment such as a load frame to use in situ neutron diffraction under stress to study the deformation behavior of materials are defined as dedicated SE equipment. Furthermore, the SE team has determined and provided guidelines at MLF for designing the instruments and SE equipment; these are called SE protocols [61]. Basically, various pieces of SE equipment belonging to individual beamlines are designed according to the guidelines. The benefit of beamline groups adopting SE protocols is that the SE team can easily support the operation and maintenance of SE equipment. Furthermore, the SE equipment can be used among several beamlines, responding to experiment contents. Available common SE equipment by category of SE is shown in Table 4 [62]. Basically, these apparatuses were introduced after consideration of MLF staff, taking into account user requests at various user meetings, etc. First of all, cryogenic and magnet SE equipment to use in both elastic and inelastic neutron scattering experiments is well supported at MLF. There are two furnaces on high-temperature SE. One of them is a niobium wire furnace that can be operated up to 1873 K.

There is SE equipment to study soft matter properties by in situ small angle neutron scattering (SANS) experiments. One is a rheometer; the other is a gas and vapor adsorption measurement instrument. These are also used in offline characterization of samples. Although there is a dedicated High Pressure Neutron Diffractometer (BL11) at MLF, versatile high-pressure SE equipment is available to use at another beamlines. As characteristic SE equipment, light irradiation SE equipment that enables us to study photo-induced phenomena during in situ experiments is also available.

Table 4. Available common sample environment (SE) equipment at MLF.

Category of SE	Available Common SE Equipment
Cryogenic and Magnet	top-loading ^{4}He cryostat 1, bottom-loading ^{3}He cryostat 1, DR insert 1, superconducting magnet 1
High Temperature	furnaces 2 (niobium/Kanthal wire)
Soft Matter	rheometer 1, gas and vapor adsorption measurement instrument 1
High Pressure	Paris-Edinburgh press 1
Light Irradiation	xenon lamp light source 1

The neutron/muon experimental facility at MLF includes two experimental halls. The SE area, which is a dedicated working space for SE equipment to develop, calibrate, and perform maintenance, is prepared in both experimental hall No. 1 and No. 2. Furthermore, there are several rooms for SE equipment in the J-PARC research building, which is a nonradioactive controlled area.

Various dedicated SE equipment belonging to individual beamlines is shown in Table 5 [62]. The first feature of the dedicated SE equipment is that whether it is elastic neutron instruments (BL08, BL15, BL17, BL18, BL20, BL21) or inelastic neutron instruments (BL01, BL02, BL12, BL14), there are many standard cryostats (refrigerators). The second feature is that there is special purpose SE equipment. There are many SE apparatuses to apply high pressure into samples in BL11. SE equipment to study deformation process is introduced in BL19, which is an Engineering Materials Diffractometer. The hydrogen absorption/desorption measurement system is available in BL21, which is a High-Intensity Total Diffractometer originally designed to study hydrogen storage materials.

Table 5. Available SE Equipment Belonging to Individual Beamline Groups.

Instrument (BL)	SE Equipment and Specifications
4SEASONS (BL01): 4D-Space Access Neutron Spectrometer	top-loading GM CCR, high-temperature stick
DNA (BL02): Biomolecular Dynamics Spectrometer	cryofurnace
iBIX (BL03): IBARAKI Biological Crystal Diffractometer	gas flow type cooling system, 3-axis goniometer
ANNRI (BL04): Accurate Neutron-Nucleus Reaction Measurement Instrument	auto sampler
NOP (BL05): Neutron Optics and Fundamental Physics	Doppler shifter, XY moving stage
SuperHRPD (BL08): Super High Resolution Powder Diffractometer	top-loading GM CCR, bottom-loading GM CCR, vanadium furnace, auto sample changer
SPICA (BL09): Special Environment Neutron Powder Diffractometer	auto sample changer
NOBORU (BL10): NeutrOn Beamline for Observation & Research Use	5-axis compact goniometer
PLANET (BL11): High Pressure Neutron Diffractometer	temperature control system (for high pressure), Paris-Edinburgh press, 6-axis multi-anvil press, pressure control system 2, vacuum chamber glove box with pressure, ruby fluorescence measurement system, Raman spectrometer
HRC (BL12): High Resolution Chopper Spectrometer	bottom-loading GM CCR, ^{3}He circulation-type refrigerator
AMATERAS (BL14): Cold-Neutron Disk-Chopper Spectrometer	top-loading GM CCR, high-temperature stick, bottom-loading GM CCR

Table 5. *Cont.*

Instrument (BL)	SE Equipment and Specifications
TAIKAN (BL15): Small and Wide Angle Neutron Scattering Instrument	sample changer, bottom-loading GM CCR 2, laser heating apparatus, electromagnet 2 (one is shared with SHARAKU)
SOFIA (BL16): Soft Interface Analyzer	laser heating stage, Langmuir trough, heater state, Peltier element stage, solid/liquid interface cell with solenoid valve option
SHARAKU (BL17): Polarized Neutron Reflectometer	4K CCR, electromagnet (shared with TAIKAN), gas-atmosphere sample cell, humidity-control cell
SENJU (BL18): Extreme Environment Single Crystal Neutron Diffractometer	bottom-loading GM CCR, bottom-loading pulse tube CCR, RT goniometer
TAKUMI (BL19): Engineering Materials Diffractometer	loading machine 3 (RT, cryogenic, high temperature), furnace system for high temperature, loading machine, 100 K cooling system for loading experiment, fatigue machine, dilatometer, Eularian cradle, Gandolfi goniometer
iMATERIA (BL20): IBARAKI Materials Design Diffractometer	auto sample changer, sample changer 2, CCR, cryofurnace, ^{3}He circulation-type refrigerator, vanadium furnace
NOVA (BL21): High Intensity Total Diffractometer	sample changer, temperature-controlled sample changer, top-loading GM CCR, high-temperature stick, hydrogen absorption/desorption measurement system
RADEN (BL22): Energy Resolved Neutron Imaging System	sample stage (large, middle, small), high temperature system

BL: Beamline; GM: Gifford-MacMahon; CCR: closed-cycle refrigerator

Detailed information about the SE equipment at MLF can be obtained from the J-PARC homepage [63]. Furthermore, users may bring their own SE equipment to use during experiments at MLF, because many kinds of experiment can be conducted due to the world-class neutron-instruments and high intensity. In this case, before the actual use of equipment, a safety examination is performed by the safety examination team, which consists of JAEA, KEK, and CROSS staff.

4.2. Development of Pulsed Magnet System

It is important to open up opportunities for new experiments that could be performed at high-intensity neutron scattering facilities such as MLF. One example is to study various quantum materials at higher magnetic fields. Therefore, a higher magnetic field device is desirable.

The magnetic field is one typical sample environment in neutron scattering experiments to study magnetic structures and magnetic excitations in various materials. Usually, steady magnetic field environments of several types, for example, horizontal magnetic field equipment and vertical magnetic field equipment, are available in neutron scattering facilities. On the other hand, in spite of the importance of the magnetic field environments, if we plan to construct a steady higher magnetic field system, usually the strengths of those magnetic field environments are limited, because there is a technical problem about high electric power consumption, which means that running costs are high. Therefore, the typical accessible strengths of such conventional magnetic field equipment are approximately more than 10 and less than 20 T. Of course, a special dedicated higher steady magnetic field system of 26 T was constructed at Helmholtz-Zentrum Berlin (HZB, Berlin-Wannsee, Germany) [64,65].

One possible solution to overcome this situation is to adopt a pulsed magnetic field environment. The pulsed magnetic sample environment for neutron diffraction was developed in the 1980s by a Japanese group [66,67]. Now, we have been developing a pulsed magnetic system to achieve up to 30 T by collaboration with the group [68]. The features of the equipment include its compactness and portability, which enable it to be used with both elastic and inelastic neutron instruments. The final goal of the system is to achieve a variable duration time and a variable magnetic field. The developed pulsed magnet system consists of three parts: a capacitor bank power supply, a small solenoid coil (see Figure 25), and a cryostat insert. We describe each component of the equipment that composes the system. Performance testing of the fully assembled system started in 2016.

4.2.1. Capacitor Bank Power Supply

The developed capacitor bank power supply is shown in Figure 25a and the specifications are listed in Table 6. The most important parameters are the pulse width of the generated current and the repetition rate. The pulse width of the current is 2.65 ms at 50% of peak; the current profile is also shown in Figure 26. The repetition rate is one pulse per several minutes, because it takes several minutes to cool the solenoid coil by liquid nitrogen. The dimensions of the housing are 760 mm width by 1000 mm length by 1410 mm height. The weight is approximately 400 kg, which allows operators to transport it between individual beamlines by themselves.

Table 6. Basic Specifications of Capacitor Bank Power Supply.

Maximum Charging Energy	Maximum Charging Voltage	Charging Time	Maximum Output Current	Pulse Width	Repetition Rate
16 kJ	2 kV	about 30 s	8 kA	2.65 ms (at 50% of peak)	one pulse per several minutes

4.2.2. Small Solenoid Coil

The developed small solenoid coil is shown in Figure 25b. The shape of the coil is a hollow cylindrical shape, and its dimensions are 33 mm outer diameter by 12 mm inner diameter by 16 mm height. The round wire of the coil is made of a high tensile strength CuAg alloy and its diameter is 1 mm. The inductance and resistance of the assembled coil are 225 μH and 71 mΩ at 77 K, respectively. The lifetime is essentially determined by the insulation breakdown of the wire covering. The insulation of the many coils does not break down more than 1000 pulses.

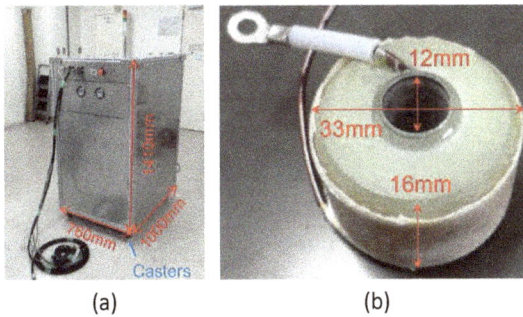

(a)   (b)

Figure 25. Components of developed pulsed magnet system: (**a**) Capacitor bank power supply; (**b**) small solenoid coil.

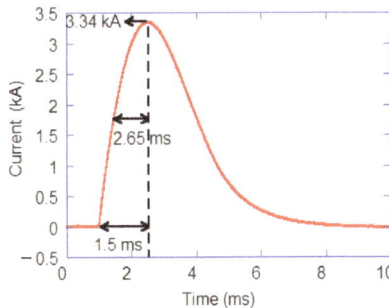

Figure 26. Current profile of the developed coil.

4.2.3. Cryostat Insert

The developed coil is inserted into a cryostat to use in neutron diffraction experiments. We fabricated the cryostat insert for the standard liquid helium flow cryostat, denoted Orange-cryostat (manufactured by AS Scientific Products Ltd, Oxfordshire, UK) with a 70 mm bore, which is one of the typical cryostats used in neutron experiments. The coil is immersed in liquid nitrogen and cooled at 77 K in order to decrease the resistance of the coil and remove the Joule heat generated by the pulsed current flow. A sample is cooled to approximately 1.5 K through helium gas. The maximum scattering angle is 30.6°. Using the developed pulsed magnetic system, test experiments were started, and it will be offered for user experiments in a couple of years.

5. Conclusions

In this paper, we have reported the developmental status of neutron devices, computational environments, and sample environments at MLF. As neutron devices, neutron detectors, optical devices including supermirror devices and ^{3}He neutron spin filters, and choppers are developed and installed at MLF. Neutron detectors used at MLF are PSDs, scintillation detectors, and gas-based two-dimensional detectors. In order to control a large number of commercially available PSDs, a readout system employing a high-speed network has been developed. Special scintillation detectors and gas-based two-dimensional detectors were also developed and installed at MLF. The supermirror devices such as neutron guides and focusing devices were developed at MLF using the IBS technique. Neutron guides with high reflectivity and large critical angle fabricated by the IBS technique were successfully installed in inelastic scattering instruments such as BL01, BL02, and BL14. In order to establish high neutron polarization using ^{3}He neutron spin filters, compact laser optics with a volume holographic grating element and a flexible non-magnetic heater have been developed for an in situ spin-exchange optical pumping application. The heater can be utilized in other applications where magnetic field disturbances are to be avoided. Developed and installed choppers at MLF are T0 choppers, Fermi choppers, and Disk choppers. These choppers are used at MLF for the purpose of background noise reduction, monochromatization of neutron beam, and elimination of undesired neutron or avoidance of frame overlap.

As for the computational environment, we have developed and equipped four software components, instrument control, data acquisition, data analysis, and a database. As instrument control software, IROHA was developed first and has now been upgraded to IROHA2. The event data recording method is employed as a standard data acquisition mode for neutron signal, conditions of sample environment equipment, status of beamline optical components, and so on. The MLF computational group has developed a software framework, Manyo Library, working on the 16 instruments for developing data reduction and analysis software. Consequently, Utsusemi software has been developed to analyze and visualize the data obtained via MLF instruments. We have also developed an integrated data management system named MLF Experimental database (MLF EXP-DB). These developments, including a remote access system are, however, still ongoing toward a higher throughput computational environment.

As for sample environment (SE) at MLF, the special SE team is organized to operate, perform maintenance, and develop common SE equipment at MLF. Consequently, a wide variety of common SE equipment is available at MLF to realize extreme sample conditions such as high and low temperatures and high magnetic fields, in addition to SE equipment belonging to individual beamline groups. This categorization enables us to expend minimum effort on operations and maintenance. The challenge of achieving a higher magnetic field is ongoing.

Acknowledgments: The authors would like to thank all MLF members who supported their development. They greatly acknowledge to former MLF division head, Yujiro Ikeda, Masatoshi Arai, and Masatoshi Futakawa for their contributions to MLF management. They also express their gratitude to Toshiji Kanaya, the present MLF division head, for his kind encouragement to write this review paper.

Author Contributions: The draft of Section 2.1 was prepared by S.S., T.S., Ka.S., T.N., K.T. and H.Y., Section 2.2 by K.So., D.Y., R.M., Section 2.3 by T.Ok., T.In., H.K., H.H. and Ke.S, Section 2.4 by S.I., K.Su, W.K., R.K., K.N., K.Sh. and M.N., Section 3 by T.Ot, T.N., Y.I., J.S., T.It., N.O. and K.M., Section 4 by K.A., S.O.-K. and M.W. Compilation of entire draft into the final version was made by Ka.S.

Conflicts of Interest: The authors declare no conflict of interest.

References

1. Satoh, S.; Muto, S.; Kaneko, N.; Uchida, T.; Tanaka, M.; Yasu, Y.; Nakayoshi, K.; Inoue, E.; Sendai, H.; Nakatani, T.; et al. Development of a readout system employing high-speed network for J-PARC. *Nucl. Instrum. Methods Phys. Res. A* **2008**, *600*, 103–106. [CrossRef]
2. Uchida, T.; Tanaka, M. Development of TCP/IP Processing Hardware. In Proceedings of the 2006 IEEE Nuclear Science Symposium Conference Record, San Diego, CA, USA, 29 October–4 November 2006; pp. 1411–1414. [CrossRef]
3. Sakasai, K.; Nakamura, T.; Katagiri, M.; Soyama, K.; Birumachi, A.; Satoh, S.; Rohdes, N.; Schooneveld, E. Development of neutron detector for engineering materials diffractometer at J-PARC. *Nucl. Instrum. Methods Phys. Res. A* **2008**, *600*, 157–160. [CrossRef]
4. Sakasai, K.; Toh, K.; Nakamura, T.; Harjo, S.; Moriai, A.; Itoh, T.; Abe, J.; Aizawa, K.; Soyama, K.; Katagiri, M.; et al. Development and Installation of Neutron Detectors for Engineering Materials Diffractometer at J-PARC. In Proceedings of the 19th Meeting on International Collaboration of Advanced Neutron Sources (ICANS-XIX), Grindelwald, Switzerland, 8–12 March 2010.
5. Nakamura, T.; Kawasaki, T.; Hosoya, T.; Toh, K.; Oikawa, K.; Sakasai, K.; Ebine, M.; Birumachi, A.; Soyama, K.; Katagiri, M. A large-area two-dimensional scintillator detector with a wavelength-shifting fibre readout for a time-of-flight single-crystal neutron diffractometer. *Nucl. Instrum. Methods Phys. Res. A* **2012**, *686*, 64–70. [CrossRef]
6. Kawasaki, T.; Nakamura, T.; Hosoya, T.; Oikawa, K.; Ohhara, T.; Kiyanagi, R.; Ebine, M.; Birumachi, A.; Sakasai, K.; Soyama, K.; et al. Detector system of the SENJU single-crystal time-of-flight neutron diffractometer at J-PARC/MLF. *Nucl. Instrum. Methods Phys. Res. A* **2014**, *735*, 444–451. [CrossRef]
7. Toh, K.; Nakamura, T.; Sakasai, K.; Soyama, K.; Hino, M.; Kitaguchi, M.; Yamagishi, H. Development of two-dimensional multiwire-type neutron detector system with individual line readout and optical signal transmission. *Nucl. Instrum. Methods Phys. Res. A* **2013**, *726*, 169–174. [CrossRef]
8. Toh, K.; Nakamura, T.; Sakasai, K.; Soyama, K.; Yamagishi, H. Evaluation of two-dimensional multiwire neutron detector with individual line readout under pulsed neutron irradiation. *J. Instrum.* **2014**, *9*, C11019. [CrossRef]
9. Mezei, F. Novel polarized neutron devices: Supermirror as spin component amplifier. *Commun. Phys.* **1976**, *1*, 81–85.
10. Soyama, K.; Ishiyama, W.; Murakami, K. Enhancement of reflectivity of multilayer neutron mirrors by ion polishing: Optimization of the ion beam parameters. *J. Phys. Chem. Solid* **1999**, *60*, 1587–1590. [CrossRef]
11. Maruyama, R.; Yamazaki, D.; Ebisawa, T.; Hino, M.; Soyama, K. Development of neutron supermirror with large-scale ion-beam sputtering instrument. *Physica B* **2006**, *385–386*, 1256–1258. [CrossRef]
12. Maruyama, R.; Yamazaki, D.; Ebisawa, T.; Hino, M.; Soyama, K. Development of neutron supermirrors with large critical angle. *Thin Solid Films* **2007**, *515*, 5704–5706. [CrossRef]
13. Wood, J. Status of supermirror research at OSMC. *Proc. SPIE* **1992**, *1738*, 22–29.
14. Maruyama, R.; Yamazaki, D.; Ebisawa, T.; Soyama, K. Development of high-reflectivity neutron supermirrors using an ion beam sputtering technique. *Nucl. Instrum. Methods Phys. Res. A* **2009**, *600*, 68–70. [CrossRef]
15. Maruyama, R.; Yamazaki, D.; Ebisawa, T.; Soyama, K. Effect of interfacial roughness correlation on diffuse scattering intensity in a neutron supermirror. *J. Appl. Phys.* **2009**, *105*, 083527. [CrossRef]
16. Kajimoto, R.; Nakajima, K.; Nakamura, M.; Soyama, K.; Yokoo, T.; Oikawa, K.; Arai, M. Study of the neutron guide design of the 4SEASONS spectrometer at J-PARC. *Nucl. Instrum. Methods Phys. Res. A* **2009**, *600*, 185–188. [CrossRef]
17. Nakajima, K.; Ohira-Kawamura, S.; Kikuchi, T.; Nakamura, M.; Kajimoto, R.; Inamura, Y.; Takahashi, N.; Aizawa, K.; Suzuya, K.; Shibata, K.; et al. AMATERAS: A Cold-Neutron Disk Chopper Spectrometer. *J. Phys. Soc. Jpn.* **2011**, *80*, SB028. [CrossRef]

18. Yamamura, K. Fabrication of Ultra Precision Optics by Numerically Controlled Local Wet Etching. *Ann. CIRP* **2007**, *56*, 541–544. [CrossRef]

19. Yamazaki, D.; Maruyama, R.; Soyama, K.; Takai, H.; Nagano, M.; Yamamura, K. Neutron beam focusing using large-m supermirrors coated on precisely-figured aspheric surfaces. *J. Phys. Conf. Ser.* **2010**, *251*, 012076. [CrossRef]

20. Yamazaki, D.; Nagano, M.; Maruyama, R.; Hayashida, H.; Soyama, K.; Yamamura, K. Neutron Focusing by a Kirkpatrick-Baez Type Super-mirror. *JPS Conf. Proc.* **2015**, *8*, 051009. [CrossRef]

21. Chupp, T.E.; Wagshul, M.E.; Coulter, K.P.; McDonald, A.B.; Happer, W. Polarized, high-density, gaseous ^{3}He targets. *Phys. Rev. C* **1987**, *36*, 2244–2251. [CrossRef]

22. Rich, D.R.; Gentile, T.R.; Smith, T.B.; Thompson, A.K. Spin exchange optical pumping at pressures near 1 bar for neutron spin filters. *Appl. Phys. Lett.* **2002**, *80*, 2210–2212. [CrossRef]

23. Moser, C.; Lawrence Ho, L.; Havermeyer, F. Self-aligned non-dispersive external cavity tunable laser. *Opt. Express* **2008**, *16*, 16691–16696. [CrossRef] [PubMed]

24. Liu, X.; Zhao, W.; Liu, H. Thermal Stress in High Power Semiconductor Lasers. In *Packaging of High Power Semiconductor Lasers*, 1st ed.; Springer: New York, NY, USA, 2015; pp. 89–105. ISBN 978-1-4939-5590-9.

25. Kissel, H.; Köhler, B.; Biesenbach, J. High-power diode laser pumps for alkali lasers (DPALs). *Proc. SPIE* **2012**, *824*, 82410Q. [CrossRef]

26. Hayashida, H.; Oku, T.; Kira, H.; Sakai, K.; Takeda, M.; Sakaguchi, Y.; Ino, T.; Shinohara, T.; Ohoyama, K.; Suzuki, J.; et al. Development and demonstration of in-situ SEOP 3He spin filter system for neutron spin analyzer on the SHARAKU polarized neutron reflectometer at J-PARC. *J. Phys. Conf. Ser.* **2014**, *528*, 012020. [CrossRef]

27. Kira, H.; Hayashida, H.; Iwase, H.; Ohishi, K.; Suzuki, J.; Oku, T.; Sakai, K.; Hiroi, K.; Takata, S.; Ino, T.; et al. Demonstration Study of Small-Angle Polarized Neutron Scattering Using Polarized ^{3}He Neutron Spin Filter. *JPS Conf. Proc.* **2015**, *8*, 036008. [CrossRef]

28. Walker, T.G. Fundamentals of Spin-Exchange Optical Pumping. *J. Phys. Conf. Ser.* **2011**, *294*, 012001. [CrossRef]

29. Babcock, E.; Nelson, I.; Kadlecek, S.; Driehuys, B.; Anderson, L.W.; Hersman, F.W.; Walker, T.G. Hybrid Spin-Exchange Optical Pumping of ^{3}He. *Phys. Rev. Lett.* **2003**, *91*, 123003. [CrossRef] [PubMed]

30. Tong, X.; Pierce, J.; Lee, W.T.; Fleenor, M.; Chen, W.C.; Jones, G.L.; Robertson, J.L. Electrical heating for SEOP-based polarized ^{3}He system. *J. Phys. Conf. Ser.* **2010**, *251*, 012087. [CrossRef]

31. Babcock, E.; Salhi, Z.; Theisselmann, T.; Starostin, D.; Schmeissner, J.; Feoktystov, A.; Mattauch, S.; Pistel, P.; Radulescu, A.; Ioffe, A. SEOP polarized ^{3}He Neutron Spin Filters for the JCNS user program. *J. Phys. Conf. Ser.* **2016**, *711*, 012008. [CrossRef]

32. Ino, T.; Hayashida, H.; Kira, H.; Oku, T.; Sakai, K. Non-magnetic flexible heaters for spin-exchange optical pumping of ^{3}He and other applications. *Rev. Sci. Instrum.* **2016**, *87*, 115108. [CrossRef] [PubMed]

33. Itoh, S.; Ueno, K.; Ohkubo, R.; Sagehashi, H.; Funahashi, Y.; Yokoo, T. T0 chopper developed at KEK. *Nucl. Instr. Methods Phys. Res. Sect. A* **2012**, *661*, 86–92. [CrossRef]

34. Kajimoto, R.; Nakamura, M.; Inamura, Y.; Mizuno, F.; Nakajima, K.; Takahashi, N.; Ohira-Kawamura, S.; Yokoo, T.; Maruyama, R.; Soyama, K.; et al. Commissioning of the Fermi-Chopper Spectrometer 4SEASONS at J-PARC—Background Study. In Proceedings of the 19th Meeting on International Collaboration of Advanced Neutron Sources (ICANS-XIX), Grindelwald, Switzerland, 8–12 March 2010.

35. Kajimoto, R.; Nakamura, M.; Inamura, Y.; Mizuno, F.; Nakajima, K.; Ohira-Kawamura, S.; Yokoo, T.; Nakatani, T.; Maruyama, R.; Soyama, K.; et al. The Fermi Chopper Spectrometer 4SEASONS at J-PARC. *J. Phys. Soc. Jpn.* **2011**, *80*, SB025. [CrossRef]

36. Fermi, E.; Marshall, J.; Marshall, L. A Thermal Neutron Velocity Selector and Its Application to the Measurement of the Cross Section of Boron. *Phys. Rev.* **1947**, *72*, 193–196. [CrossRef]

37. Itoh, S.; Ueno, K.; Yokoo, T. Fermi chopper developed at KEK. *Nucl. Instrum. Methods Phys. Res. A* **2012**, *661*, 58–63. [CrossRef]

38. Nakamura, M.; Nakajima, K.; Kajimoto, R.; Arai, M. Utilization of multiple incident energies on Cold-Neutron Disk-Chopper Spectrometer at J-PARC. *J. Neutron Res.* **2007**, *15*, 31–37. [CrossRef]

39. Nakamura, M.; Kajimoto, R.; Inamura, Y.; Mizuno, F.; Fujita, M.; Yokoo, T.; Arai, M. First Demonstration of Novel Method for Inelastic Neutron Scattering Measurement Utilizing Multiple Incident Energies. *J. Phys. Soc. Jpn.* **2009**, *78*, 093002. [CrossRef]

40. Nakamura, M.; Kajimoto, R. General Formulae for the Optimized Design of Fermi Chopper Spectrometer. *JPS Conf. Proc.* **2014**, *1*, 014018. [CrossRef]
41. Suzuki, J.; Murakami, K.; Manabe, A.; Kawabata, S.; Otomo, T.; Furusaka, M. Object-oriented data analysis environment for neutron scattering. *Nucl. Instrum. Methods Phys. Res. A* **2004**, *534*, 175–179. [CrossRef]
42. Yasu, Y.; Nakayoshi, K.; Sendai, H.; Inoue, E.; Tanaka, M.; Suzuki, S.; Satoh, S.; Muto, S.; Otomo, T.; Nakatani, T.; et al. Development of DAQ-Middleware. In Proceedings of the 17th International Conference on Computing in High Energy and Nuclear Physics (CHEP09), Prague, Czech Republic, 21–27 March 2009; p. 022025. [CrossRef]
43. Mcgreevy, R.; Otomo, T.; Anderson, I.; Miller, S.; Geist, A. New opportunities for data analysis software: An international Collaboration. *Neutron News* **2004**, *15*, 25–27. [CrossRef]
44. Nakatani, T.; Inamura, Y.; Ito, T.; Harjo, S.; Kajimoto, R.; Arai, M.; Ohhara, T.; Nakagawa, H.; Aoyagi, T.; Otomo, T.; et al. The Implementation of the Software Framework in J-PARC/MLF. In Proceedings of the 12th International Conference on Accelerator and Large Experimental Physics Control Systems, Kobe, Japan, 12–16 October 2009; p. 673.
45. Nakatani, T.; Inamura, Y.; Ito, T.; Otomo, T. Data acquisition and device control software framework in MLF, J-PARC. In Proceedings of the 21st Meeting of the International Collaboration on Advanced Neutron Sources (ICANS-XXI), Mito, Japan, 29 September–3 October 2014; p. 493. [CrossRef]
46. Nakatani, T.; Inamura, Y.; Ito, T.; Otomo, T. The Control Software Framework of the Web Base. In Proceedings of the 2nd International Symposium on Science at J-PARC, Tsukuba, Japan, 12–15 July 2014; p. 036013. [CrossRef]
47. Nakatani, T.; Inamura, Y.; Ito, T.; Moriyama, K. IROHA2: Standard instrument control software framework in MLF, J-PARC. In Proceedings of the New Opportunities for Better User Group Software NOBUGS2016, Copenhagen, Denmark, 17–19 October 2016; p. 76. [CrossRef]
48. Nakatani, T.; Inamura, Y.; Ito, T.; Otomo, T.; Satoh, S.; Muto, S.; Nakayoshi, K.; Sendai, H.; Inoue, E.; Yasu, Y. Event Mode Data Acquisition System at MLF/J-PARC. In Proceedings of the 19th Meeting on International Collaboration of Advanced Neutron Sources (ICANS-XIX), Grindelwald, Switzerland, 8–12 March 2010.
49. Nakatani, T.; Inamura, Y.; Moriyama, K.; Ito, T.; Muto, S.; Otomo, T. Event recording data acquisition system and experiment data management system for neutron experiments at MLF, J-PARC. In Proceedings of the 12th Asia Pacific Physics Conference (APPC12), Chiba, Japan, 14–19 July 2013; p. 014010. [CrossRef]
50. Uchida, T. Hardware-Based TCP Processor for Gigabit Ethernet. *IEEE Trans. Nucl. Sci.* **2008**, *55*, 1631–1637. [CrossRef]
51. Nakayoshi, K.; Yasu, Y.; Inoue, E.; Sendai, H.; Tanaka, M.; Satoh, S.; Muto, S.; Kaneko, N.; Otomo, T.; Nakatani, T.; et al. Development of a Data Acquisition Sub-System using DAQ-Middleware. *Nucl. Instrum. Methods Phys. Res. A* **2009**, *600*, 173–175. [CrossRef]
52. Nakayoshi, K.; Yasu, Y.; Inoue, E.; Sendai, H.; Tanaka, M.; Satoh, S.; Muto, S.; Suzuki, J.; Otomo, T.; Nakatani, T.; et al. DAQ-Middleware for MLF/J-PARC. *Nucl. Instrum. Methods Phys. Res. A* **2010**, *623*, 537–539. [CrossRef]
53. Yasu, Y.; Nakayoshi, K.; Sendai, H.; Inoue, E. Functionally of DAQ-Middleware. *IEEE Trans. Nucl. Sci.* **2010**, *57*, 487–490. [CrossRef]
54. Suzuki, J.; Nakatani, T.; Ohhara, T.; Inamura, Y.; Yonemura, M.; Morishima, T.; Aoyagi, T.; Manabe, A.; Otomo, T. Object-oriented data analysis framework for neutron scattering experiments. *Nucl. Instrum. Methods Phys. Res. A* **2009**, *600*, 123–125. [CrossRef]
55. Suzuki, J.; Inamura, Y.; Ito, T.; Nakatani, T.; Otomo, T. "Manyo-Lib" Object-Oriented Data Analysis Framework for Neutron Scattering. In Proceedings of the New Opportunities for Better User Group Software NOBUGS2016, Copenhagen, Denmark, 17–19 October 2016; p. 72.
56. Inamura, Y.; Nakatani, T.; Suzuki, J.; Otomo, T. Development status of software 'Utsusemi' for Chopper Spectrometers at MLF, J-PARC. *J. Phys. Soc. Jpn.* **2013**, *82*, SA031. [CrossRef]
57. Harjo, S.; Ito, T.; Aizawa, K.; Arima, H.; Abe, J.; Moriai, A.; Iwahashi, T.; Kamiyama, K. Current Status of Engineering Materials Diffractometer at J-PARC. *Mater. Sci. Forum* **2011**, *681*, 443–448. [CrossRef]
58. Kawasaki, T.; Ito, T.; Inamura, Y.; Nakatani, T.; Harjo, S.; Gong, W.; Iwahashi, T.; Aizawa, K. Neutron Diffraction Study of Piezoelectric Material under Cyclic Electric Field using Event Recording Technique. In Proceedings of the 21st Meeting of the International Collaboration on Advanced Neutron Sources (ICANS-XXI), Mito, Japan, 29 September–3 October 2014; pp. 528–531. [CrossRef]

59. Moriyama, K.; Nakatani, T. A Data Management Infrastructure for Neutron Scattering Experiments in J-PARC/MLF. In Proceedings of the 15th International Conference on Accelerator and Large Experimental Physics Control Systems ICALEPCS2015, Melbourne, Australia, 17–23 October 2015; p. 834.

60. Moriyama, K.; Nakatani, T. Recent Progress in the Development of MLF EXP-DB in J-PARC. In Proceedings of the New Opportunities for Better User Group Software NOBUGS2016, Copenhagen, Denmark, 17–19 October 2016; p. 80. [CrossRef]

61. Aso, T.; Yamauchi, Y.; Sakaguchi, Y.; Munakata, K.; Ishikado, M.; Ohira-Kawamura, S.; Yokoo, T.; Watanabe, M.; Takata, S.; Hattori, T.; et al. Present status of sample environment at J-PARC MLF. In Proceedings of the 21th International Collaboration on Advanced Neutron Source (ICANS-XXI), Mito, Japan, 29 September–3 October 2014.

62. Ohira-Kawamura, S.; Oku, T.; Watanabe, M.; Takahashi, R.; Munakata, K.; Sakaguchi, Y.; Ishikado, M.; Ohuchi, K.; Hattori, T.; Kira, H.; et al. Sample Environment at the J-PARC MLF. *J. Neutron Res.* **2017**, *19*, 15–22. [CrossRef]

63. J-PARC. Available online: http://j-parc.jp/researcher/MatLife/en/se/equipment.html (accessed on 27 July 2017).

64. Steiner, M.; Tennant, D.A.; Smeibidl, P. New high field magnet for neutron scattering at Hahn-Meitner Institute. *J. Phys. Conf. Ser.* **2006**, *51*, 470–474. [CrossRef]

65. Smeibidl, P.; Tennant, A.; Ehmler, H.; Bird, M. Neutron Scattering at Highest Magnetic Fields at the Helmholtz Centre Berlin. *J. Low Temp. Phys.* **2010**, *159*, 402–405. [CrossRef]

66. Nojiri, H.; Takahashi, K.; Fukuda, T.; Fujita, M.; Arai, M.; Motokawa, M. 25T repeating pulsed magnetic fields system for neutron diffraction experiments. *Physica B* **1998**, *241–243*, 210–212.

67. Nojiri, M.; Motokawa, H.; Takahashi, K.; Arai, M. 30 T repeating pulsed field system for neutron diffraction. *IEEE Trans. Appl. Supercond.* **2000**, *10*, 534–537. [CrossRef]

68. Watanabe, M.; Nojiri, H.; Itoh, S.; Ohira-Kawamura, S.; Kihara, T.; Masuda, T.; Sahara, T.; Soda, M.; Takahashi, R. Development of compact high field pulsed magnet system for new sample environment equipment at MLF in J-PARC. In Proceedings of the International Symposium of Quantum Beam Science at Ibaraki University, Mito, Japan, 18–20 November 2016.

quantum beam science

MDPI

Review

Materials and Life Science Experimental Facility at the Japan Proton Accelerator Research Complex IV: The Muon Facility

Wataru Higemoto [1], **Ryosuke Kadono** [2,*], **Naritoshi Kawamura** [2], **Akihiro Koda** [2],
Kenji M. Kojima [2], **Shunsuke Makimura** [2], **Shiro Matoba** [2], **Yasuhiro Miyake** [2],
Koichiro Shimomura [2] **and Patrick Strasser** [2]

[1] J-PARC Center and Advanced Science Research Center, JAEA, Tokai, Ibaraki 319-1195, Japan;
     higemoto.wataru@jaea.go.jp
[2] J-PARC Center and Institute of Materials Structure Science,
     High Energy Accelerator Research Organization (KEK), Tsukuba, Ibaraki 305-0801, Japan;
     nari.kawamura@kek.jp (N.K.); coda@post.kek.jp (A.K.); kenji.kojima@kek.jp (K.M.K.);
     shunsuke.makimura@kek.jp (S.Mak.); shiro.matoba@kek.jp (S.Mat.); yasuhiro.miyake@kek.jp (Y.M.);
     koichiro.shimomura@kek.jp (K.S.); patrick.strasser@kek.jp (P.S.)
*   Correspondence: ryosuke.kadono@kek.jp; Tel.: +81-29-284-4896

Academic Editor: Klaus-Dieter Liss
Received: 1 May 2017; Accepted: 8 June 2017; Published: 15 June 2017

Abstract: A muon experimental facility, known as the Muon Science Establishment (MUSE), is one of the user facilities at the Japan Proton Accelerator Research Complex, along with those for neutrons, hadrons, and neutrinos. The MUSE facility is integrated into the Materials and Life Science Facility building in which a high-energy proton beam that is shared with a neutron experiment facility delivers a variety of muon beams for research covering diverse scientific fields. In this review, we present the current status of MUSE, which is still in the process of being developed into its fully fledged form.

Keywords: positive muon; muonium; muon spin rotation; negative muon; muonic atom; muonic X-ray

1. Overview of the Muon Science Establishment

1.1. Muon Science Facility with Tandem-Type Production Graphite Target

The Materials and Life Science Facility (MLF) consists of neutron and muon science facilities that utilize a 3-GeV proton beam (beam power 1 MW, repetition rate 25 Hz). It was decided to use the muon production target in tandem with that for neutrons on a single proton beamline (see Figure 1) rather than constructing one for muons in a separate building with a dedicated proton beamline and beam dump. This resulted in the significant reduction of the total construction cost by sharing the beam between the neutron and muon facilities, thereby enabling the common use of utilities while avoiding the enormous work of beam dump construction, including high-level tritium water handling. The total beam loss induced by placing the muon production target is specified to be 10% or less. Although two muon-production targets made of graphite with thicknesses of 10 mm and 20 mm could be installed in the beamline upstream from the neutron target [1], as was originally planned, a single 20-mm-thick graphite target is currently utilized for muon production. The heat deposited into a 25-mm-diameter spot on this target by a 1-MW proton beam is estimated to be 3.9 kW.

In the earlier stage of development, an edge-cooled fixed (non-rotating) graphite target was adopted, because of the relative ease of handling and maintaining it. Since September 2008, the fixed graphite target has been utilized without any trouble. In 2014, to reduce the frequency of future target replacement due to the relatively short lifetime expected for the fixed target in the 1-MW era, a rotating graphite target based on a working model developed at the Paul Scherrer Institute (PSI) was installed [2].

1.2. Building Structure and Maintenance

The MLF building consists of proton beamline tunnels (M1 and M2) and two wings for experimental halls (numbers 1 and 2, on the east and west sides, respectively). The tunnel structure is intended to contain radioactive products within the tunnel that may be emitted during maintenance of the neutron and/or muon targets. Since a certain fraction of the primary 3-GeV proton beam is scattered toward the neutron target, two sets of beam collimators (called "scrapers") are installed to prevent radiation damage to the beamline components, including the quadrupole magnets and beam ducts. Based on the experiences accumulated at the PSI in dealing with 1-MW proton beams, all of the beamline components were designed to enable remote handling from the maintenance area located 2.4 m above the beamline level [2].

1.3. Muon Beamlines

In Phase 1 of the original Japan Proton Accelerator Research Complex (J-PARC) construction plan, which was split into two parts, a superconducting decay/surface muon beamline (D-line) with a modest acceptance (about 40 msr) for pion capture was installed in hall No. 2, where both surface and decay muons can be delivered to experimental areas D1 and D2. A total surface muon (μ^+) flux of 1.8×10^6/s was achieved using a 120-kW proton beam in 2009, surpassing the previous world-record flux realized at the RIKEN muon facility at the ISIS facility of the Rutherford Appleton Laboratory (RAL) in the UK [3]. A total surface muon flux of 1.5×10^7/s is anticipated to be achievable by utilizing the designed proton beam power of 1 MW.

Following the successful operation of the D-line, the "Super Omega" muon beamline (U-line) was installed, which consists of a large-acceptance rad-hard solenoid, a superconducting curved transport solenoid, and superconducting axial focusing magnets, which together produce an "ultra-slow muon" (USM) beam.

In experimental hall No. 1, a new surface muon beamline (S-line) dedicated to the material sciences and a "high-momentum" muon beamline (H-line) are under construction. The S-line is planned to have four experimental areas (S1–S4), and the beamline to area S1 was completed by 2016.

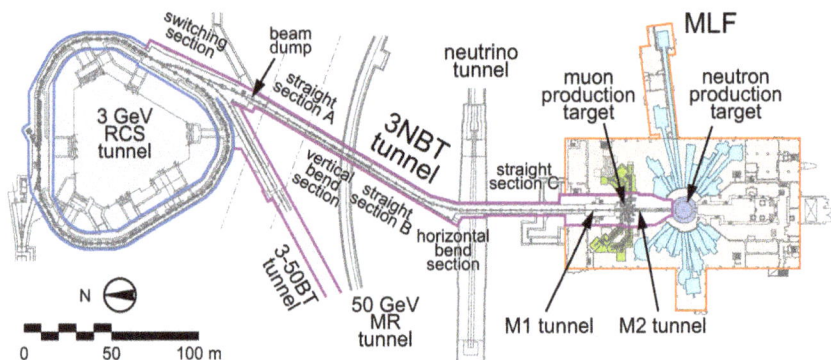

Figure 1. Schematic drawing of the proton beamline from the 3-GeV rapid cycling synchrotron (RCS) to the Materials and Life Science Facility (MLF).

2. Muon Target

2.1. Overview of the Muon Target

As shown in Figure 1, the proton-beamline tunnel runs through the center of the MLF building from north to south. The muon production target is located about 30 m upstream from the neutron production target. To prevent the diffusion of radioactive contamination generated around the muon target, the proton beamline tunnel is isolated from the experimental halls. The muon target system and front-end devices for the secondary muon beamlines are located in the proton beamline tunnel (M2). The muon target system consists of the target itself and the adjacent beam scrapers that prevent protons scattered by the target from reaching the downstream devices between the muon and neutron targets and causing radiation damage. Some of the components near the muon target are expected to be highly radioactive, making it impossible to perform hands-on maintenance. In the case of the muon target, the radiation dose rate is predicted to be 5 Sv/h after a year of 1-MW proton beam irradiation. Thus, these components are designed to be handled remotely from the maintenance area located 2.4 m above the beamline level and separated by 2-m-thick Fe shielding.

As described previously in Section 1.3, the highest flux pulsed muon beam in the world is produced by a 20-mm-thick graphite target (IG-430U, Toyotanso Co., Ltd., Osaka, Japan) [4], in which about 5% of a proton beam is spared for muon production. More specifically, 4% is lost due to reactions with C nuclei and 1% is for Coulomb scattering. After muons were first produced in September 2008, the proton beam intensity was gradually increased to 300 kW in January 2013 for stable operation [5,6]. The fixed muon target remained in service without replacement until May 2014.

Proton irradiation causes radiation damage to the material properties of graphite [7,8]. In particular, the effects on the dimensions of graphite are serious. The lifetime of the fixed target was estimated to be 1 year, based on a simulation of graphite irradiation with a 1-MW proton beam. Because the muon target is highly activated by proton irradiation, the used target must be handled remotely in a "hot cell", which requires considerable time, cost, and manpower [9]. Thus, it is crucial to extend the target lifetime to minimize the frequency with which the facility operation must be interrupted to replace it. To this end, the development of a rotating target, in which the radiation damage is spread across a wider area, began in 2008, simultaneously with the operation of the fixed target [10]. Since its installation into the proton beamline in September 2014, the rotating muon target has operated successfully without replacement.

2.2. Fixed Muon Target

The fixed muon target is made of 20-mm-thick, 70-mm-diameter isotropic graphite. The beam profile has a Gaussian distribution with a standard deviation of approximately 3.5 mm. The energy deposited by the 1-MW proton beam on the muon target has been estimated to be 3.9 kW using a particle transport simulation code (PHITS) [11,12]. Because the target is located in a vacuum chamber, the cooling water traveling through the stainless-steel tube must remove most of the heat embedded in the Cu frame. To absorb the thermal stress, a Ti layer is employed as an intermediate material between the graphite target and Cu frame [13]. Figure 2 shows a picture of the fixed muon target, taken from 12 m upstream along the proton beamline, where the thermocouples measuring the temperature of the Cu frame can be seen. Upon replacement of the fixed target with the rotating target in 2014, the fixed target had incurred no damage during proton beam operation for nearly 6 years.

Figure 2. Picture of the fixed muon target, taken from 12 m upstream along the proton beamline (before upgrading to the current rotating target).

Radioactive damage induced by proton irradiation deteriorates the material properties of graphite, such as its thermal conductivity, strength, and dimensions. Dimensional changes of graphite could have particularly serious effects. These factors must be addressed when evaluating the target lifetime by considering three kinds of stresses. One type of stress is the residual (mostly compressive) stress induced by shrink-fitting during manufacturing to maintain high thermal conduction at the interfaces of different materials. The second type is thermal stress due to the temperature distribution caused by proton beam irradiation, where compressive stress is expected due to the restriction of the thermal expansion of the high-temperature beam spot by the surrounding low-temperature part. The thermal conductivity degradation due to proton irradiation must be considered when simulating this type of stress. The third type of stress is the tensile stress resulting from the dimensional changes after continuous beam irradiation for a certain time. The highly irradiated beam spot on the graphite disk will exhibit shrinkage under the constraints imposed by the surrounding part, where the accumulated stress is proportional to the operational time and hysteresis.

The simulation indicated that the theoretical lifetime of the graphite target was less than 1 year under 1-MW proton irradiation. To evaluate the practical lifetime of the currently used target, the detailed proton irradiation history, including the beam position on the target, which was altered at each operational time to disperse the irradiation damage, was carefully considered. The beam spot was concentrically moved away from the center in 4-mm increments to 8 different positions almost every 1000 h when the beam power was 200 kW and every 700 h when the beam power was 300 kW [14].

The fixed muon target was successfully replaced with the new rotating target in September 2014 [15]. After temporary storage for cooling in a storage pod, the highly radioactive used target was contained within a shielding vessel called a "transfer cask" for transport to the hot cell, where it was handled and replaced with the new target using the remote handling devices. The used target was cut into pieces by a special device in the hot cell to enable the storage volume available for the highly activated parts to be used as efficiently as possible.

Remote handling commissioning in the hot cell has been conducted almost once a year. Figure 3 shows pictures of the target chamber in which the muon target and stainless-steel shielding are inserted, the transfer cask during the replacement of the fixed target with the rotating target, and a mockup target loaded onto a remote-handling device in the hot cell, from left to right.

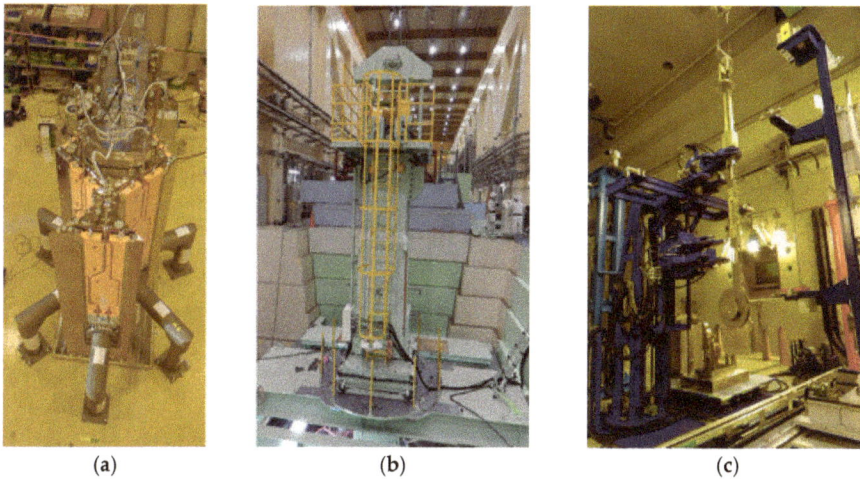

Figure 3. Pictures of (**a**) the target chamber, (**b**) the transfer cask, and (**c**) remote handling commissioning in the hot cell.

2.3. Rotating Muon Target

The new rotating target is designed to have a horizontal shaft parallel to the proton beam line to serve as a rotation axis. Because the motor used to rotate the target must be located 2.4 m above the beamline level to avoid exposure to high radiation doses, the rotating motion is transmitted to the horizontal shaft through a long vertical shaft and pair of bevel gears. Figure 4a shows a picture of the entire target assembly. As shown in Figure 4b, the rotating body is composed of a graphite wheel, a wheel support, and the horizontal shaft supported by two horizontal bearings. The two bearings are attached to a cooling jacket in which water piping is embedded. The temperatures of the graphite wheel and hottest bearing were estimated to be 940 K and 390 K, respectively. The target rotation speed was determined to be 15 rpm based on evaluation of the maximum temperature gradients inside the graphite wheel.

Because the lifetime of the graphite wheel under continuous rotation is estimated to be more than 30 years, the bearing lifetime will be the factor limiting the lifetime of the entire system. The bearings (supplied by JTEKT Co., Ltd., Osaka, Japan) [16] are located in a high vacuum of 10^{-4} Pa, at a high temperature of 390 K, and under a high radiation dose of 100 MGy/year. In these conditions, disulfide molybdenum (MoS_2) or Ag is generally used as a coating lubricant. We adopted a sintered *compact* of disulfide tungsten (WS_2) instead, because WS_2 coating is expected to have a much longer lifetime than MoS_2 or Ag coating (~110,000 h, or 22 years assuming proton beam operation for 5000 h/year).

Since the fixed target was replaced with the rotating target, no problems have developed over a period including 300–400 kW proton beam operation for 3 months, 500-kW operation for 1 month, and 600-kW operation for 1 h. Approximately 1×10^{22} protons have been supplied to the rotating target, and the cumulative number of revolutions has reached 4.8 million.

During proton beam operation, the temperatures of the cooling jacket, horizontal shaft, and graphite are monitored. Because it is difficult to measure the temperatures of the rotating bodies directly, the temperature increases in the thermocouples due to thermal radiation are monitored. The temperature of the horizontal shaft is continuously monitored using thermocouples inserted into a hollow core in the center of the shaft. The thermocouples for target monitoring are isolated by thermal shielding and are irradiated only by the graphite disk. While the measured temperatures of the horizontal shaft were found to agree closely with the simulated results, those of the graphite disk exhibited a significant discrepancy whose cause is yet to be clarified. In any case, the effects of heat generation due to proton beam irradiation must be carefully confirmed during actual operation. Table 1 compares the measured and simulated temperatures at typical beam powers.

Figure 4. Pictures of (a) the rotating muon target assembly; and (b) the rotating muon target itself.

Table 1. Measured and simulated temperatures of the horizontal shaft and maximum simulated temperatures of the graphite at typical beam powers.

	300 kW	500 kW	1 MW
Shaft Simulation	71 °C	84 °C	115 °C
Shaft Measurement	78 °C	95 °C	-
Graphite Simulation	400 °C	475 °C	620 °C

2.4. Scrapers

Two scrapers (SCs, made of O-free Cu) are placed downstream from the rotating muon target in a series to reduce the proton beam halo. Each of them is approximately 300 mm wide, 700 mm high, and 700 mm long in the direction of the beam axis and has a circular bore along the beam axis with an inlet diameter of 74 mm that increases downstream. They are exposed to heat loads of about 19 kW during 1-MW operation due to the beam halo induced by proton scattering from the target. The cooling water pipes are embedded in the SCs using the hot isostatic press technique to minimize the thermal resistance at the interfaces between the pipes and Cu block. K-type thermocouples enclosed in stainless steel sheaths are mounted on the upstream and side surfaces to measure the SC temperatures.

We encountered a problem in that the SC temperatures measured by the thermocouples increased unexpectedly after upgrading to the rotating target. It was later determined that the local heating of the halo monitor mounted on the upstream surface of each SC was causing thermal radiation to the thermocouples that were directly facing hot spots. Therefore, the SCs were replaced with improved SCs without halo monitors, where thermocouples were attached behind the boring structure as shown in Figure 5, so that they would not be directly exposed to the thermal radiation from the upstream components, including the rotating target. For the replacement work, the radioactivity induced by the beam irradiation was evaluated using PHITS and a residual radioactivity calculation code, DCHAIN-SP. The radioactivity due to all of the radionuclides was determined to be about 20 TBq.

As of 2017, the new SCs are running smoothly with 150 kW operation. The temperature of the new SCs was measured to be 42 °C during 500-kW operation in the second half of 2015, which is close to the design value, suggesting that temperature measurement of the new SCs should be feasible even during 2-MW operation.

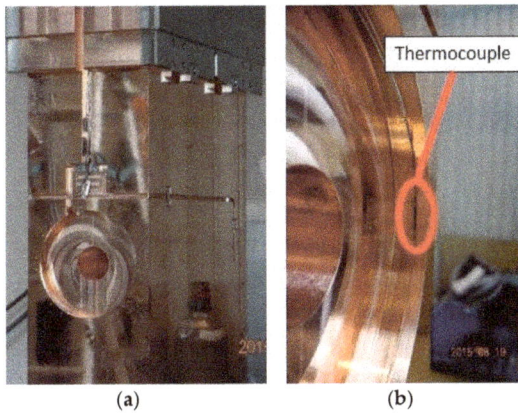

(a)   (b)

Figure 5. Pictures of (**a**) one of the new scrapers (SCs); and (**b**) the region near the thermocouple.

3. Muon Beamlines

Muons are emitted by pions generated via nuclear reactions in the graphite target, where two different modes are commonly used to extract them as secondary beams. The first type of secondary beam, called a "surface-muon" beam, is produced by collecting the muons emitted by the positive pions that come to a halt at the subsurface of the graphite target. The other type is called a "decay-muon" beam and is produced by collecting the muons from positive or negative pions that decay during flight (i.e., transport) along the beamline.

As shown in Figure 6, the muon-production target is surrounded by four muon beamlines (the D-, U-, S-, and H-lines) extending to the two experimental halls. The beamlines are designed to provide high-flux muon beams with different properties to meet the demands of a variety of muon experiments. Two of the beamlines form 60° angles with respect to the proton beamline (forward), and the others are at 135° (backward). The front-end parts of these beamlines are located within the proton beamline tunnel, leaving little room for future modification as a trade-off for radiation safety. The beamline construction proceeded successively from the D-line, to the U-line, and then to the S-line, which have been in operation since 2009, 2011, and 2015, respectively, while the H-line is still under construction.

The D-line can deliver both negatively and positively charged muons with momenta of up to 120 MeV/c, in addition to surface muons (~30 MeV/c) and so-called "cloud muons," where the latter are the muons generated by the lowest energy ("cloud"-like) pions emitted by the muon target. This high flexibility in terms of beam momentum/charge is suitable for serving a wide variety of user programs. The U-line is next to the D-line in hall No. 1 and is characterized by high muon acceptance, enabling the delivery of the highest flux surface-muon beam among the four secondary beamlines. This high-flux beam is used to generate a USM beam via the resonant ionization of thermal muonium by a laser [17]. The S-line is designed to transport surface muons to four experimental areas in experimental hall No. 2 to provide beams simultaneously to muon spin rotation (μSR) spectrometers using muon-kicker devices. The construction is currently complete up to the first experimental area (S1). The H-line is designed to meet the demands of particle physics experiments that require long-term occupancy of beamlines/experimental areas and therefore are incompatible with short-term user programs for the material sciences.

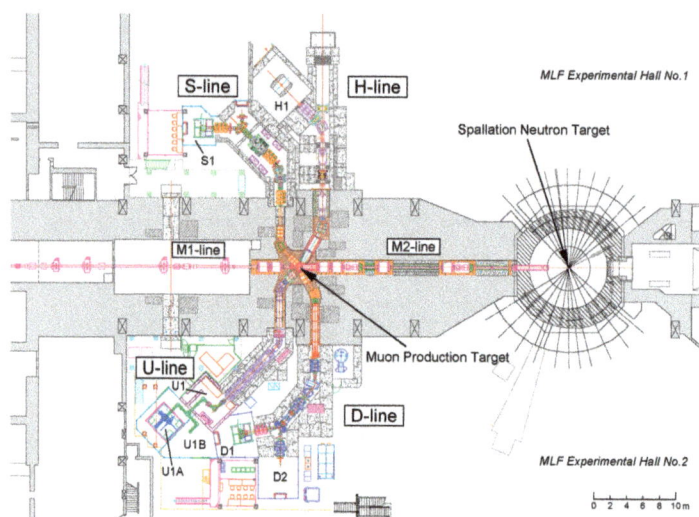

Figure 6. Floor plan of the muon beamlines (provisional for the H-line) in the MLF building. Four beamline tunnels, one each for the D-, U-, S-, and H-lines, surround the muon production target.

3.1. D-Line

The D-line is the first muon beamline to have been completed in the Muon Science Establishment (MUSE) facility [18], and it is currently employed for various kinds of muon experiments, including condensed matter physics and material science experiments using the μSR technique, nondestructive elemental analysis, and fundamental muon physics investigations. It can deliver surface (positive) muons with a total flux of 10^7–10^8/s and decay positive/negative muons with momenta of up to 120 MeV/c and a flux of 10^6–10^7/s at 60 MeV/c. Recently, a total surface muon flux of 4.5×10^6/s over a beam spot size of ø40 mm FWHM at the sample position was achieved during 300 kW operation. As shown in Figure 7, the entire beamline consists of three main sections: one for pion injection, a superconducting decay solenoid, and one for muon extraction. It has two branches downstream from the last bending magnet ("septum" magnet), each of which extends to an experimental area (D1 or D2). A DC separator ("Wien filter") placed after the second bend is used in combination with slits to separate muons from the positrons/electrons in the beam.

Similarly to the proton beam produced at ISIS-RAL, the beam provided by the rapid cycling synchrotron (RCS) at J-PARC has a double-pulsed structure in which each pulse has a width of ~100 ns; the pulses are separated by 600 ns and are delivered at a repetition rate of 25 Hz. Accordingly, the muon beam has nearly the same time structure as that at ISIS-RAL. To utilize the muon beam more efficiently, a muon-kicker system was installed to split the beam into two single-pulsed beams for simultaneous delivery to the two experimental areas, enabling two μSR experiments to be conducted in parallel [19]. The system consists of two magnetic kickers, two switchyard magnets, and a septum magnet. The switchyard magnets deflect the first muon pulse of a double-pulsed beam so that the beam is shifted to the right at the entrance of the septum magnet to be transported to area D1. Then, the kicker magnets deflect the second muon pulse in the opposite direction to inject it into the septum magnet on the left that leads to area D2. The rise time for kicker excitation is less than 300 ns with a flat top of about 300 ns, while the decay time is not critical since the following muon pulse arrives only after 40 ms. The system was designed for operation with a momentum of up to 60 MeV/c to deliver beams simultaneously to areas D1 and D2, while a double-pulsed beam can be delivered alternately to one of these areas with a momentum of up to 120 MeV/c.

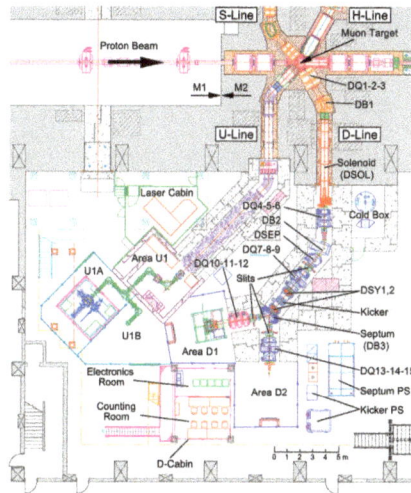

Figure 7. Layout of the D-line in MLF experimental hall No. 2. The key components are the superconducting decay solenoid that brings muons from the proton beam tunnel into the experimental hall, the kicker system that separates the double-pulsed muon beam, and the septum magnet that delivers muons to areas D1 and D2.

Until recently, the D-line relied on a superconducting solenoid magnet that had been in service for more than 30 years since its first installation at the former Booster Meson facility at KEK. It had a cold bore with an inner radius of ø12 cm and was equipped with thermal shield windows at both ends that prevented the transport of relatively low-momentum muons due to scattering at those windows. The solenoid was finally replaced in 2015 by a new one containing an inner warm bore with an inner radius of ø20 cm and no thermal windows. This replacement enabled the extraction of intense negative muons with momenta much lower than those that could be extracted previously. Preliminary beam commissioning results showed that there was more than a tenfold increase in the intensity of negative muons with momenta less than 20 MeV/c. This new development will be highly advantageous for experiments involving nondestructive elemental analysis of thin samples.

3.2. U-Line

The main purpose of the U-line is to deliver a high-flux surface muon beam for the production of intense USMs, where "ultra-slow" means a kinetic energy of 10^2–10^3 eV, which is low enough to be stopped within a few nanometers from the surfaces of materials. Such muons will extend the scope of the μSR technique from bulk materials to thin films, near-surfaces, interfaces, and extremely small samples and will facilitate not only a wide variety of nano-science studies, but also novel 3D imaging with "ultra-slow muon microscopes" [20].

USMs are produced as a tertiary beam by the re-acceleration of thermal muons regenerated by the laser resonant ionization of muonium atoms evaporated from a hot W foil, a method that originated at the Meson Science Laboratory at KEK [21]. Since the yield of USMs relative to that of surface muons is ~10^{-3}, the muon flux available at the end of the U-line is crucial for the feasibility of the present method based on hot W foil.

The U-line, or "Super-Omega muon beamline," consists of three magnet devices, i.e., a normally conducting capture solenoid, a superconducting curved transport solenoid, and an axial focusing solenoid [22,23]. Muons with momenta of up to 45 MeV/c can be captured with a large acceptance solid angle (400 msr) of the front-end solenoid in tunnel M2, which faces the upstream side of the muon target at 135°. The captured muons are then transported to the experimental hall by a superconducting curved transport solenoid and to area U1 by an axial focusing solenoid. Since the beamline consists entirely of solenoids, surface μ^+ and cloud μ^- are transported simultaneously, which is inconvenient for most experiments. Muon charge selection is performed using two dipole coils installed in the straight section of the curved solenoid [24]. Recently, we achieved the highest pulsed muon flux in the world with a time-averaged surface-muon intensity of 6.4×10^7/s using a proton beam power of 212 kW, which is 20 times more intense than at the D-line and is equivalent to 3.0×10^8/s with the designed beam power of 1 MW.

The layout of the USM beamline is illustrated in Figure 8. In area U1, a high-flux muon beam irradiates a hot W foil placed in the muonium chamber, and laser beams passing near the foil surface ionize muonium atoms evaporated into the vacuum. To ionize the muonium atoms efficiently, a resonant ionization scheme involving the "1s–2p–unbound" transition and a pulsed nanosecond laser was adopted. A new state-of-the-art all solid-state laser system produces vacuum ultraviolet (VUV) light with a wavelength of 122.088 nm (Lyman-alpha), which is necessary to induce the 1s–2p transition [25,26]. After laser ionization, the free muons have a mean thermal kinetic energy of only 0.2 eV with 50% spin polarization. They are re-accelerated up to 30 keV and focused by an electrostatic lens for transport via a series of electric quadrupoles and electric bends to two experimental areas, U1A (for USM-μSR) and U1B (for further re-acceleration, microbeam μSR, muon transmission microscopy, etc.) [27]. In area U1A, a μSR spectrometer furnished with a load-lock system for sample exchange in an ultra-high vacuum is installed on a high-voltage platform ($\pm$30 kV) to enable USM implantation without causing discharge around the sample (see Section 4.2). During this stage of commissioning, a total USM flux of 42 μ^+/s has been attained (February 2017), which is higher than that achieved previously at RAL (20 μ^+/s). This flux was measured at an intermediate position after the first electric bend with a beam size of approximately 10 mm × 15 mm (FWHM). Further improvements are in progress to enhance the laser power and efficiency of beam transportation to the μSR spectrometer, where a final beam size of a few millimeters is expected. After completion of the initial commissioning phase, experiments will be conducted in area U1A.

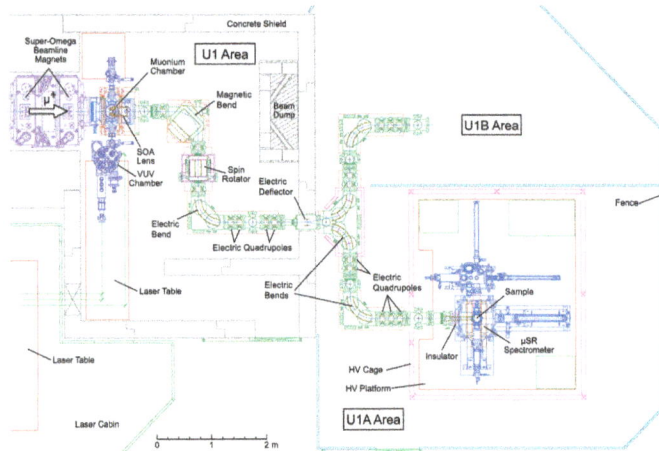

Figure 8. Layout of the ultra-slow muon (USM) beamline. A high-flux surface muon beam is injected into a hot W target in the muonium chamber (left), from which thermal muonium atoms evaporate into a vacuum. These muonium atoms are ionized by a laser system and accelerated and focused by an electrostatic lens for transport at 30 keV. The transport beamline consists of one magnet bend, three electric bends, and electric quadrupoles, and it is electrostatically isolated from the high-voltage cage accommodating the muon spin rotation (μSR) spectrometer.

3.3. S-Line

The S-line, which is located in experimental hall No. 1 of the MLF building, is designed to provide surface-muon beams (with momenta of 28 MeV/c, which are called "slow" muon beams) for conventional μSR experiments simultaneously performed in the four experimental areas. The beamline is designed to explore the potential advantages of high-flux surface muons in condensed matter physics, and it is typically expected that it will facilitate the investigation of smaller samples and/or require shorter data acquisition (DAQ) periods. In particular, the latter characteristic enables stroboscopic observations of time-evolving phenomena to be performed. While the beamline will eventually serve four experimental areas (S1–S4, as indicated in Figure 9), so far only the branch to area S1 has been completed. The construction work performed thus far includes that for the power supply yard and concrete shield, which were completed in 2013, followed by the installation of beamline magnets and other components in 2014. The first delivery of surface muons to area S1 was confirmed by time-of-flight measurements conducted during the beamline commissioning performed in 2015.

The basic concept of the S-line is similar to that of the D-line, which enabled duplicate design of some components, such as the vacuum chamber and DC separator, including the high-voltage electrodes and correction magnets that are currently used in the D-line. Meanwhile, a Cockcroft–Walton-type high-voltage generator, which had served the U-line DC separators and proved to be more stable than the corresponding device in the D-line, was adopted to achieve stable beam operation. Furthermore, the DC separator is located upstream from the second bending magnet, unlike in the D-line. It is expected that the drift length between the DC separator and the beam slits, which is longer than that in the D-line, will provide improved separation and reduce positron contamination.

Figure 9. Current layout of the S-line to experimental area S1 (**left**) and a provisional floor plan for the completed S-line complex with four branches, one to each of the experimental areas (**right**). The cross-hatched areas near the shielding wall to the M1/M2 tunnel indicate the power supply yards.

The main difference between the S- and D-lines is that an *electric* kicker system is used to deliver single-pulsed muons to area S1 (and to areas S2–S4 in the future). In particular, symmetric operation to distribute two pulsed beams to areas S1 and S2 within a switching time of less than 300 ns is realized using bipolar high-voltage power supplies based on solid-state Marx generators (Figure 10). In addition, a window-frame-type switchyard magnet is included in the kicker chamber to utilize double-pulsed muon beams. It should be noted that the influence of leakage fields from the septum magnet to the beam orbit toward areas S3 and S4 is not negligible. To minimize this influence, magnetic stainless steel is employed for the septum vacuum chamber. Every power supply for the beamline magnets is equipped with a polarity changer, so that negative muons (cloud muons) can be transported to area S1.

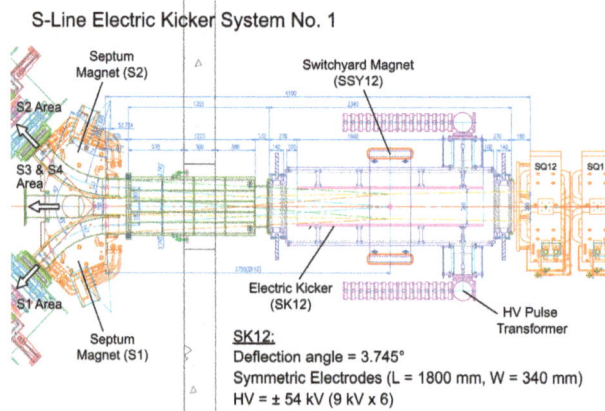

Figure 10. Layout of the S-line electric kicker system that delivers single-pulsed muon beam to areas S1 and S2. The muon beam can also be sent straight to areas S3 and S4, where installation of a similar kicker system is planned.

Since the first beam delivery to area S1, beamline commissioning has been intensively performed using the μSR spectrometer, ARTEMIS (the Advanced Research Targeted Experimental Muon Instrument at S-line; see Section 4.1), sponsored by the Element Strategy Initiative Project for Electronic Materials. A custom-made automatic beam-tuning program is used to obtain a well-focused muon beam at the sample position by monitoring μ-e decay events. A typical surface muon flux of 8.5×10^4/s for a tailored beam spot size of ø20 mm at the sample position has been obtained at 150 kW operation. The commissioning processes for the spectrometer itself and the associated sample environments, such as those related to cryostat operation, sample temperature logging, and unmanned measurements, including DAQ control, are performed in parallel. The IROHA2 framework, which is the standard protocol at MLF for communicating among devices to control sample environments, went online in area D1 soon after a brief test in area S1. Not only the software, but also the hardware, such as the water-cooled thermal shield used to stabilize the detector temperature, was exported after testing in area S1. It should be stressed that area S1 plays an important role in the seamless upgrading of the μSR equipment.

3.4. H-Line

While the H-line was originally designed to produce a "high-momentum" muon beam, the experiments currently under consideration rather require high muon flux as well as momentum tunability [28–30], which will also be important for future experiments. To meet these demands, a new beamline optics concept involving a large-aperture muon-capture solenoid, wide-gap bending magnet, and pair of two solenoid magnets with oppositely directed fields was proposed. For the detailed design of the beam optics, however, conventional matrix calculations are not applicable due to the failure of the near-axis approximation for a large-aperture solenoid. Thus, Monte-Carlo particle tracking simulation code, G4BEAMLINE [31], was applied to optimize the beamline magnet parameters. Figure 11 depicts typical surface-muon beam transport results.

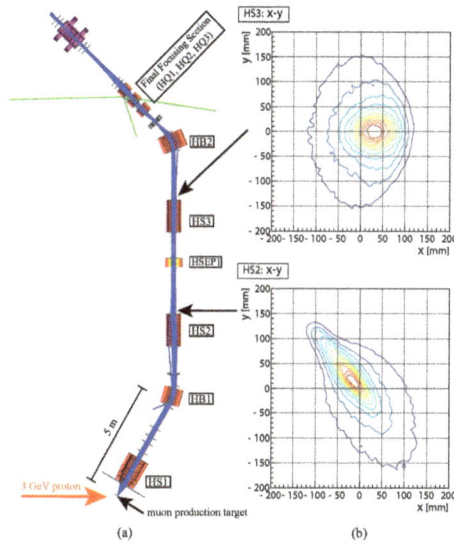

Figure 11. (a) Typical results obtained by the Monte-Carlo particle tracking simulation code, G4BEAMLINE. The blue lines are the beam trajectories. A surface muon beam from a point source is transported to the first experimental area through three solenoid magnets, HS1, HS2 and HS3, and two bending magnets, HB1 and HB2. The Wien filter is also depicted, although it is not in use in this transmission. (b) The beam profiles at the exits of the second and third solenoid magnets, HS2 and HS3, are provided.

The front-end part of the M1/M2 proton beam tunnel has been completed, and the remaining downstream part is under construction, which should enable early completion of J-PARC Phase 1 and will provide beams for two experiments, i.e., high-precision measurement of the hyperfine splitting of muonium [28] and searches for muon-to-electron conversion [29]. In Phase 2, the H-line will be extended beyond the east wall of the MLF building to accommodate a muon beam accelerator with a USM source by employing a technique similar to that used in the U-line and will eventually be utilized for precision measurements of the anomalous magnetic moment of muons [30] and transmission muon microscopy.

The expected total surface-muon flux for the H-line is about 10^8/s. The high flux attainable using large-aperture magnets has a trade-off relationship with duct-streaming neutrons, which would be more serious in the H-line than in the other muon beamlines. The radiation shield and interlock devices are designed to ensure safe operation against such background radiation.

4. Muon Experiment Instruments

4.1. μSR Spectrometers

μSR is an experimental technique that is used to measure the time evolution of muon spin polarization in condensed matter. The μSR spectrometer consists of detectors for positrons/electrons emitted by muons, whose spatial asymmetry carries information about the muon spin polarization; electronic devices to measure the time between the muon entry and positron decay times; and an apparatus to control the sample environment characteristics, such as the magnetic field and temperature. In this section, the detectors and electronic devices in the common μSR spectrometer structure are described first and are followed by details about the sample environment apparatus.

It is difficult to apply μSR to pulsed muon sources, such as J-PARC, due to the enormous rate of positron decay at each beam pulse. As explained in the Section 1.3, J-PARC provides the highest flux of pulsed muons in the world; one pulse contains $\sim 10^5$ muons, which decay with a mean lifetime of 2.2 μs. Thus, the instantaneous positron count rate can be as high as 100 Gcps. Since the typical response time of a fast positron counter (plastic scintillator) is 10 ns, the positron counters must be divided into thousands of channels of small scintillators so that each individual detector measures a reduced count rate on the order of 100 Mcps to handle such high event rates. Thus, a pulsed μSR spectrometer must contain thousands of independent detectors, whose time histograms must be stored independently. To this end, we developed a scalable many-channel positron detector system called Kalliope for use in μSR experiments.

Kalliope (KEK Advanced Linear and Logic-board Integrated Optical detectors for Positrons and Electrons) is an all-in-one detector system for time differential measurement. One Kalliope detector unit is depicted in Figure 12 and consists of 32 channels of individual positron detectors. The time differences between the common start trigger pulse and the multiple hit events on the detectors are recorded by a time-to-digital converter with 1-ns timing resolution and stored in memory on a digital board for subsequent transfer to the DAQ PC via an Ethernet cable using TCP/IP (Transmission Control Protocol/Internet Protocol). The amplifier and discriminator for signals from plastic scintillators (with avalanche photo-diodes (APDs) for photon detection) with a digitally controlled threshold level are realized in an integrated circuit on the analog board, which is configured by the firmware on the digital board. Although Kalliope is compact, it contains everything necessary to measure decay positrons from muons; with Kalliope, Ethernet hubs, and a DAQ-PC, one can measure μSR without preparing other electronics, which represents a major advantage over the conventional NIM (Nuclear Instrumentation Module) and CAMAC (Computer Automated Measurement and Control) or VMEbus (Versa Module Europa bus) crate-based electronics.

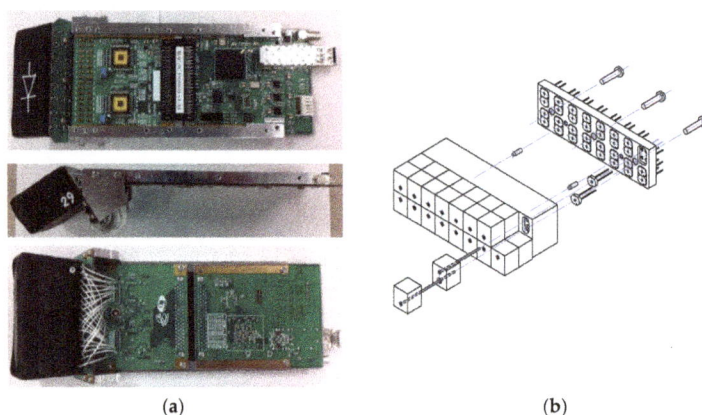

(a)   (b)

Figure 12. (**a**) One Kalliope (KEK Advanced Linear and Logic-board Integrated Optical detectors for Positrons and Electrons) detector, which consists of a scintillator block on a scinti-board, an analog board, and a digital board. Data are transferred to a data acquisition (DAQ) PC via an Ethernet cable. (**b**) A scintillator block consisting of 32 scintillator channels (with dimensions of 10 mm × 10 mm × 12 mm), each with wavelength-shifting fibers and pixel-type avalanche photo diodes (MPPC by Hamamatsu) installed.

Two identical general-purpose μSR spectrometers are installed in areas S1 and D1 (see Figure 13). The spectrometer in area S1 is named ARTEMIS (hereafter denoted as S1-ARTEMIS). Each of these spectrometers contains 40 Kalliope units that cover 21.2% of the total solid angle for positron/electron detection. The detector arrangement is illustrated in Figure 14. The power supply and Ethernet hubs for the Kalliope units are stored in the base of each spectrometer for portability. The largest Helmholtz coils can apply magnetic fields of up to 0.4 T at the sample position with a DC current of 1000 A. The magnet base has a rotating table with a stopper pin, so that the 0.4 T field can be applied using either longitudinal or horizontal transverse field geometry. Each spectrometer is also equipped with vertical square Helmholtz coils (14 mT for 100 A) and three sets of stray-field compensation coils (1 mT for 10 A) to achieve a zero field at the sample position. The specifications of these spectrometers are summarized in Table 2.

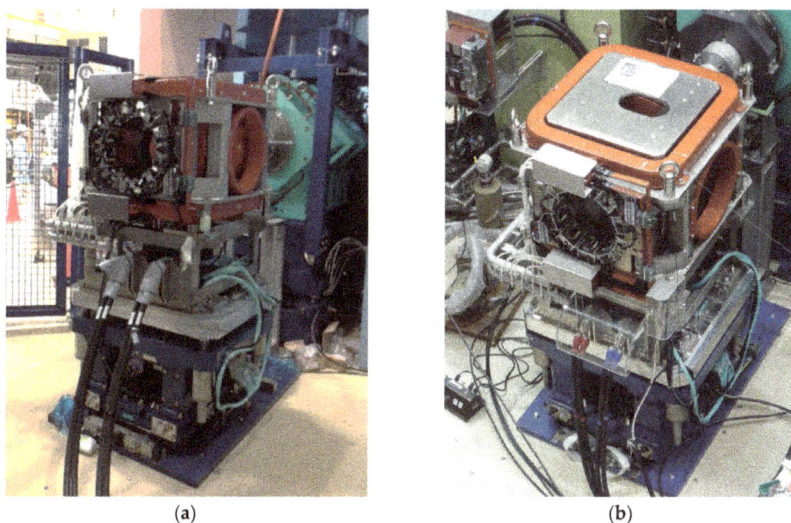

(a) (b)

Figure 13. Twin general-purpose μSR spectrometers: (**a**) S1-Advanced Research Targeted Experimental Muon Instrument at S-line (S1-ARTEMIS); and (**b**) the D1 instrument. Each spectrometer includes 40 Kalliope units for positron/electron detection that cover 21.2% of the solid angle. Magnets are equipped to apply magnetic fields of up to 0.4 T and 14 mT in the horizontal and vertical directions, respectively, and xyz stray-field compensation coils are employed to produce a field of 0 T at the sample position.

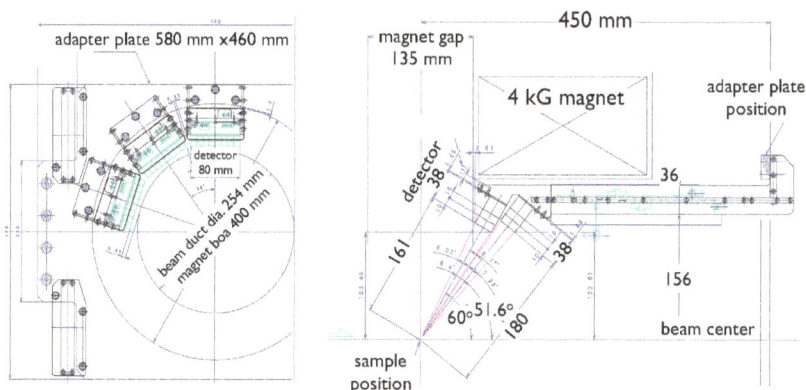

Figure 14. Arrangements of the positron/electron counters in the twin general-purpose spectrometers.

Table 2. Specifications of the twin µSR spectrometers in areas D1 and S1.

Magnets	
Horizontal field B_Z (//beam) or B_x ($\perp$beam)	Maximum 0.4 T (1000 A, 100 V)
Bore and gap	ø400, 135 mm
Homogeneity	0.1% ø20 mm, 1% ø50 mm
Stability	Drift < 0.01% for 8 h Ripple Vp-p < 0.01%
Vertical field B_y (($\perp$beam, vertical)	Maximum 14 mT (25 mT with optional coils) (100 A, 40 V)
Homogeneity	0.1% ø20 mm, 1% ø50 mm
Stability	Drift < 0.05% Ripple Vp-p < 0.1%
Correction field coils used to achieve a zero field (CCx, y, z)	Maximum 1 mT (10 A, 20 V)
Homogeneity	0.5% ø20 mm, 10% ø50 mm
Positron Counters	
Number of channels	32 channels × 40 Kalliope units = 1280 channels (640 pairs)
Detector solid angle	21.2% of 4π (2 × 8 array of 10 mm × 12 mm scintillator cubes: covers an area of 1920 mm² at 180 mm and 161 mm from the sample)
Sample Insert Area	
Bore and gap	ø254, 135 mm

Because of the instantaneous high count rate of positrons/electrons in pulsed-µSR measurements, the transient properties of the preamplifiers that are related to the signals from the APDs embedded in the custom analog chip (application-specific integrated circuit, ASIC) are essential to ensure the high performances of the Kalliope units, which significant efforts have been made to improve. The latest ASIC (called "Volume 2014") consists of 100 MHz current amplifiers with a pole-zero cancelation circuit for time constant conversion. Figure 15 presents the example of µSR spectra obtained using S1-ARTEMIS with a count rate of 200 Mhits/h, which corresponds to the rate expected for a 16 mm × 16 mm sample and 1 MW proton beam power.

The computer interface for user operation of the µSR spectrometer is based on the IROHA2 framework developed at MLF, a common user interface for DAQ and device control furnished with auto-run features. The sample environment characteristics, such as the temperature and magnetic field, are set using a PC in the auto-run sequence, and their stability is waited for if necessary before the measurement run starts. The run ends automatically when the preset count is achieved, and the sequence proceeds to the next step. The auto-run progress and current sample environment status are continuously displayed on a web page, so that the user can monitor the experiment remotely via the Internet.

For conventional µSR measurements, various sample environments are available. For the sample temperature, a He-free top-loading dilution refrigerator (≥50 mK), a He gas-flow cryostat (3–400 K), and an infrared furnace (≤1000 K) are ready for use. As mentioned above, magnetic fields can be applied along the beam direction (≤0.4 T) or perpendicular to the beam direction (≤0.025 T). It is also possible to perform µSR measurements under light illumination using a flush lamp synchronized with the muon pulses.

To enable µSR measurements to be performed on small samples (≥5 mm × 5 mm) by reducing the background events from muons that missed the sample, a "fly-past chamber" is used. The fly-past chamber is a long vacuum vessel in which samples are suspended with minimal surrounding equipment for cryogenic control, and muons that missed the sample fly away downstream to prevent the positrons/electrons that they emit from hitting the detectors.

To improve the time resolution, which is limited by the beam pulse width of 80–100 ns, a device called a "muon beam slicer" was once installed in the D-line to area D1 [32]. The beam slicer consists of an electric kicker, a pulsed electric power supply with a fast rise time, and a correction magnet. By applying a pulsed electric field (±80–100 kV) that is synchronized with the muon pulses, a muon pulse can be sliced to have a narrow structure (≥20 ns) at the expense of reducing the muon flux.

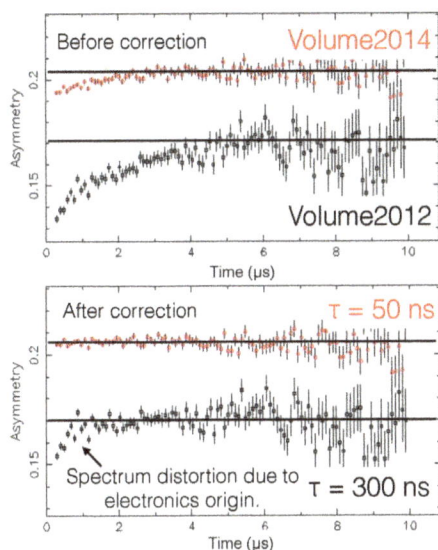

Figure 15. µSR time spectra obtained by S1-ARTEMIS with the earlier application-specific integrated circuit (ASIC) ("Volume2012") and the latest ASIC ("Volume2014"). The top and bottom panels respectively depict the spectra before and after pileup correction using the standard formula $N = N_{obs}/(1 - \tau N_{obs})$, where τ is the dead-time parameter. (The Volume2012 spectra suffer from distortion due to the insufficient transient performances of the voltage amplifiers.)

4.2. µSR Spectrometer for USMs

USMs with energies of ~0.2 eV are immediately accelerated to 30 keV for transport to the µSR spectrometer at the end of the USM beamline installed in area U1A (Figure 16). The entire spectrometer assembly, which includes a cryostat, Kalliope positron detectors, magnets, other electronic devices, and their power supplies, is contained within a large metallic cage on an electrically isolated stage. The beam implantation energy can be varied from 0.2 kV to 30 kV by tuning the electrostatic potential of the stage to decelerate USMs in the entry section of the spectrometer. The corresponding muon implantation depth range is 0–200 nm for Cu, where the specimen properties can be investigated continuously from near the surface to the bulk region.

The required features of the USM-µSR spectrometer are identical to those of conventional spectrometers, except that the sample environment that must be compatible with ultra-high vacuum conditions. Meanwhile, care must be taken to minimize the thermal radiation to control the sample temperature without beam windows or radiation shields, which would interrupt the USM implantation. The spectrometer is furnished with a set of Helmholtz coils for applying external magnetic fields of up to 0.14 T along the beam direction. Since a muon spin rotator (Wien filter) is installed upstream from the spectrometer along the USM beamline to modify the initial muon spin direction from 0° (along the beam direction) to 90°, one can perform µSR measurements under transverse or longitudinal field conditions using a single set of magnets. To enable sample exchange without requiring the high vacuum to be broken, a load-lock chamber is attached next to the main vacuum chamber of the spectrometer. The load-lock chamber is also used for sample preparation and surface condition characterization.

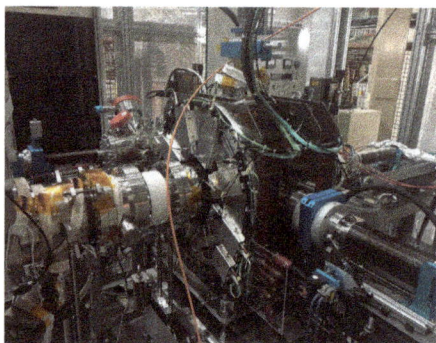

Figure 16. μSR spectrometer installed in area U1A for use in USM experiments. The entire assembly is placed in a large cage on a stage electrically isolated from the ground.

5. Recent Muon Application Highlights

5.1. μSR Studies in Condensed Matter Physics and the Materials Sciences

μSR is one of the most natural muon applications in that it utilizes the sensitivity of magnetic fields to explore the electronic properties of matter via implanted muons. As implanted "pseudo-hydrogen," muons also provide unique opportunities to probe the simulated state of interstitial H via the muon-electron hyperfine parameters. Because of the negligibly small difference (~0.5%) between the reduced electron masses of H and neutral muonium (Mu^0, the muonic analogue of a neutral H^0 atom), the electronic state of implanted Mu^0 is mostly equivalent to that of interstitial H^0.

As mentioned in Section 4.1, areas D1 and S1 in MUSE are currently furnished with general-purpose μSR spectrometers, whose abilities to be used for conventional μSR measurements are almost identical. Although the sample environment characteristics, e.g., the accessible temperature and magnetic field ranges, are limited so far, these instruments have proven the usefulness of μSR for the investigation of the local electronic properties of a variety of materials.

One such example is the identification of a new antiferromagnetic phase in a prototype Fe-based superconductor, $LaFeAsO_{1-x}H_x$ (LFAO-H), over a range of unprecedentedly high carrier concentrations [33]. It was previously reported that the range of carrier doping could be extended to much higher concentrations ($x > 0.2$) by substituting O with H instead of F, and the secondary maximum of the superconducting transition temperature (T_c) was observed with $x \sim 0.35$ [34]. The results of a preliminary nuclear magnetic resonance (NMR) study suggested the emergence of a certain kind of magnetism in the overdoped region [35]. Based on this finding, a team of experts on muons, neutrons, and synchrotron radiation (SR)-X-rays (working together for the "Element Strategy Initiative" project being conducted at the Condensed Matter Research Center) has made a concerted effort with a group from Tokyo Institute of Technology to elucidate the electronic properties of LFAO-H with large H concentrations. They employed the muon spin rotation technique to map out the dependence of the Néel temperature (T_N) on x in a timely fashion (Figure 17). Subsequent neutron and SR-X-ray measurements indicated that the magnetic structure is different when x is high compared to when it is low and that the magnetic transition accompanies a structural change to a non-centrosymmetric structure (Aem2) at $x = 0.5$. These observations suggest that the new phase might be regarded as another "parent" for the secondary superconducting phase.

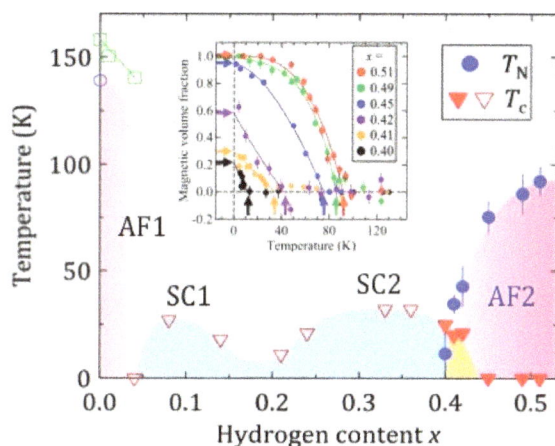

Figure 17. New magnetic phase (AF2), for which the Néel temperature (T_N) was mapped out by performing μSR measurements. Inset: Magnetic volume fraction vs. temperature determined via μSR [33]. T_c: superconducting transition temperature.

Yet another example is the observation of a peculiar magnetic ground state in an Ir spinel compound. Geometrical frustration in electronic degrees of freedom, such as spin, charge, and orbit, which is often realized in stages in highly symmetric crystals, has been an important topic in condensed matter physics. Inorganic compounds with AB_2X_4 spinel structures have provided platforms for the investigation of unusual physical properties related to geometrical frustration. A thiospinel compound, $CuIr_2S_4$, is a recent example of such a compound. In $CuIr_2S_4$, mixed-valent Ir ions form isomorphic octamers, $Ir^{3+}_8S_{24}$ and $Ir^{4+}_8S_{24}$, with lattice dimerization of Ir^{4+} pairs in the latter, upon the occurrence of a metal–insulator (MI) transition at 230 K. While it has been proposed that a pair of Ir^{4+} atoms ($5d^5$, total spin $S = 1/2$) forms a non-magnetic spin-singlet dimer driven by orbital order and associated Peierls instability, the magnetic properties of the ground state remain to be clarified microscopically.

The results of recent muon and Cu-NMR studies have indicated that a spin glass-like *magnetic* ground state can be realized in $CuIr_2S_4$ below ~100 K [36]. As shown in Figure 18, slow Gaussian damping was observed at 200 K, which is expected for muons exposed to random local magnetic fields from nuclear magnetic moments (primarily from ^{63}Cu and ^{65}Cu nuclei in $CuIr_2S_4$), indicating that the compound is non-magnetic at this temperature. In contrast, fast exponential depolarization sets at below ~100 K, where the depolarization rate as well as the relative amplitude of the depolarizing component increases with decreasing temperature. These observations contradict the naive expectation based on the currently accepted scenario that Ir^{4+} pairs form non-magnetic spin-singlet states upon MI transition. The spin glass-like behavior suggests that competing interactions influence the Ir^{4+} atoms, leading to a magnetically frustrated state. It has been proposed that the spin-orbit interaction, which has gained rapid recognition in recent years as an important factor in the physics of 5d electron systems, e.g., Sr_2IrO_4, might be the origin of this frustration as well as unquenched local spins, where Ir^{4+} is represented as an eigenstate of an effective isospin $J_{eff} = 1/2$ multiplet.

It is known that substitution of Cu with Zn in $CuIr_2S_4$ suppresses the MI transition, eventually leading to superconductivity in $Cu_{1-x}Zn_xIr_2S_4$ for $x > 0.25$. A μSR study of Zn-substituted samples indicated that the spin glass-like magnetism was strongly suppressed compared with that in $CuIr_2S_4$ (see Figure 18c,d), which parallels the characteristics of high-T_c cuprates and/or Fe-pnictides. Thus, $CuIr_2S_4$ may serve as a new platform for the study of superconductivity under the strong influence of the spin-orbit interaction.

Meanwhile, muons are used to simulate the electronic structure of interstitial H in materials in which H may play a significant role. One related topic that recently gained prominence was the study of the electronic properties of barium titanate (BaTiO$_3$). The perovskite oxide BaTiO$_3$ is one of the most important ferroelectric materials that is widely used in electronic devices. It exhibits ferroelectricity at ambient temperature, and the associated high dielectric constant is indispensable for downsizing multilayer ceramic capacitors.

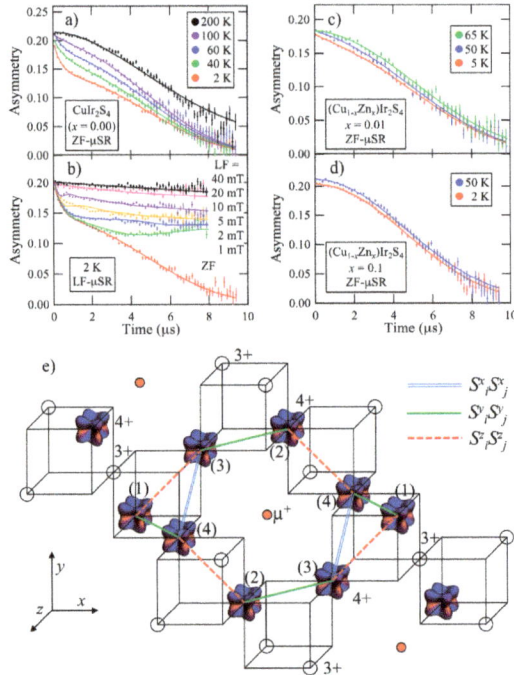

Figure 18. (**a**) Time-dependent μSR spectra observed at several temperatures in a powder sample of CuIr$_2$S$_4$ under zero external field. (**b**) μSR spectra at 2 K under various longitudinal fields. μSR spectra in Cu$_{1-x}$Zn$_x$Ir$_2$S$_4$ under zero external field with Zn contents of (**c**) x = 0.01 and (**d**) x = 0.1. (**e**) Octamer configuration associated with charge order in CuIr$_2$S$_4$. The exchange interaction $S^\gamma_i S^\gamma_j$ for each Ir^{4+} pair is shown by a line along the relevant γγ bond (γ = x, y, z). The octupolar manifold at each Ir^{4+} site represents the spin density profile in a hole with an isospin-up state (an eigenstate of the J_{eff} = 1/2 multiplet under strong spin-orbit interaction). The bond lengths of the (1)–(4) and (2)–(3) Ir^{4+} pairs are reported to shrink by 15% upon charge ordering and the associated structural transition [36].

Infrared absorption spectroscopy results have indicated that H impurities in BaTiO$_3$ may form O–H bonds, suggesting that H can be stabilized as interstitial H$^+$ to form OH$^-$ ions. First-principles calculation results have further suggested that the electronic levels associated with the OH$^-$ state may not be formed in the band gap, remaining near the bottom of the conduction band so that it can serve as a shallow electron donor. The carrier doping by OH$^-$ formation may cause a serious problem, specifically, that the performance of BaTiO$_3$ as an insulating material for capacitors could be degraded by H impurities in the environment.

A μSR experiment was conducted to test this possibility using muons to simulate the electronic state of interstitial H in matter [37]. Satellite signals were observed around the center line (corresponding to the μ$^+$ state) in the μSR spectra measured in BaTiO$_3$ at low temperatures, which is a typical sign that Mu^0 with an extremely small hyperfine parameter is formed (see Figure 19a).

The Mu^0 state was also found to disappear upon warming up to a temperature above 100 K, which is interpreted as "ionization" of Mu^0 occurring with a small activation energy of ~10 meV (see Figure 19b). Thus, muonium has been demonstrated to act as a shallow donor in $BaTiO_3$, strongly suggesting that interstitial H would exhibit similar behavior.

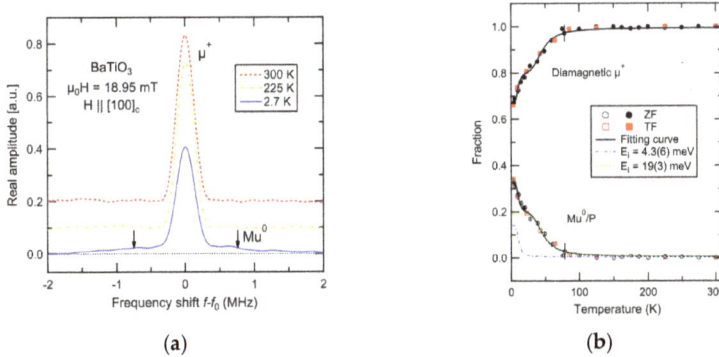

Figure 19. (**a**) Fourier transform of the μSR spectrum of $BaTiO_3$. (**b**) Temperature dependences of the fractional yields of μ^+ and Mu^0 [37].

5.2. Non-Invasive Element Analysis

When a negative muon is implanted into matter, it loses kinetic energy, is captured by the Coulomb field of the nucleus, and forms a muonic atom consisting of the nucleus and a negatively charged muon in place of an electron. As illustrated in Figure 20, upon muonic atom formation in a highly excited level, such as that with a principal quantum number of ~15, the captured muon immediately de-excites to lower muon atomic levels by emitting characteristic muonic X-rays with energies unique to each element in the matter. This feature enables the use of muonic X-rays for non-destructive element analysis.

Because the mass of a muon is large (207 times that of an electron), muonic X-rays are about 200 times harder (i.e., higher in energy) than electronic X-rays from the same element. For Cu, for instance, the muonic Kα X-ray has an energy of about 1500 keV, whereas that of the electronic X-ray is only 8 keV. While the latter is blocked by 100-μm-thick Cu foil, muonic X-rays can easily penetrate a Cu plate several millimeters thick. Thus, one of the major advantages of muonic X-rays over electronic X-rays is that they can be used for element analysis deep inside a bulk specimen. This capability is particularly useful for light elements, since conventional fluorescent X-ray analysis is insensitive to elements lighter than Na due to the limited penetration depths of X-rays. An example of a muonic X-ray spectrum obtained for a meteorite specimen is shown in Figure 20, where signals from light elements are detected through a wall of glass tube [38].

Moreover, since tuning the beamline parameters can vary the implantation energy of negative muons, they can be stopped at any depth from a few micrometers to several centimeters. Therefore, muonic X-rays provide a unique means of 3D bulk-elemental analysis when combined with a 2D X-ray imaging device or beam-scanning technique.

As mentioned in Section 3.1 the negative muons delivered at the D-line originate from π^- decays in flight in a long superconducting solenoid magnet. Although the beamline was originally designed to transport muons with energy of up to ~165 MeV (corresponding to a momentum of 250 MeV/c), the specifications of the current beamline components only enable the transport of muons with momenta of up to 120 MeV/c. Another boundary on the low-energy side was caused by the beam windows of the superconducting solenoid that are placed at both ends to protect the vacuum and low temperature within the cold bore.

Recently, we installed a new superconducting solenoid containing a warm bore with an inner radius of ø20 mm without beam windows so that negative muons with extremely low momenta can be delivered to area D2. Beam commissioning is currently in progress to attain a relatively high flux and narrow momentum width to expand the scope of muonic X-ray element analysis using low-energy negative muons.

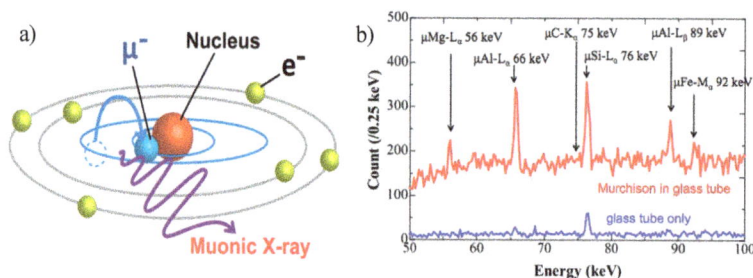

Figure 20. (**a**) Schematic illustration of the formation of a muonic atom, an atomic system that has a negatively charged muon in place of an electron. The element-specific X-ray is emitted upon successive transitions of muons to the lower-energy muonic orbitals. (**b**) Muonic X-ray spectra from the powdered Murchison meteorite in a SiO_2 glass tube. A clear signal of Mg and a marginally resolved signal of Fe from the sample were detected through the 1-mm thick SiO_2 glass wall [38].

Acknowledgments: We would like to thank all of the members of the MUSE construction team and collaborators, in particular, Jack L. Beveridge, Jaap Doornbos, Gerd Heidenreich, Yoshiro Irie, Yasuhiro Makida, Kazutaka Nakahara, Toru Ogitsu, Naohito Saito, and Soshi Takeshita, for their significant contributions during the early stage. Thanks also to the present and past MUSE staff, including Taihei Adachi, Hiroshi Fujimori, Yutaka Ikedo, Takashi U. Ito, Yasuo Kobayashi, Jumpei Nakamura, Kaneta Nagamine, Takashi Nagatomo, Yu Oishi, and Amba D. Pant, for their crucial contributions in the development of the MUSE facility. We also would like to express our sincere gratitude to Shoji Nagamiya and Yujiro Ikeda for the kind support provided by J-PARC Center under their directorship. Finally, we dedicate this article as a special tribute to the late Kusuo Nishiyama, who devoted his entire life to bringing MUSE into existence.

Author Contributions: The draft of Section 1 was prepared by Y.M., Section 2 by S.Mak. and S. Mat, Section 3 by K.S., P.S., A.K., and N.K., Section 4 by K.M.K and W.H., Section 5 by R.K. and Y.M., respectively according to their expertizes and primary fields of contribution to the MUSE facility. Revision of the entire draft for integration into the final production was made by R.K.

Conflicts of Interest: The authors declare no conflict of interest.

References

1. Miyake, Y.; Nishiyama, K.; Kawamura, N.; Makimura, S.; Strasser, P.; Shimomura, K.; Beveridge, J.L.; Kadono, R.; Fukuchi, K.; Sato, N.; et al. Status of J-PARC muon science facility at the year of 2005. *Physica B* **2006**, *374*, 484–487. [CrossRef]
2. Heidenreich, G.; Baumann, P.; Geissler, A.; Strinning, A.; Wagner, W. Improvement of the operational reliability of target-E. *PSI Sci. Rep.* **1998**, *VI*, 16.
3. Nagamine, K.; Matsuzaki, T.; Ishida, K.; Watanabe, I.; Nakamura, S.N.; Kadono, R.; Kawamura, N.; Sakamoto, S.; Iwasaki, M.; Tanase, M.; et al. New RIKEN-RAL pulsed μCF facility and X-ray studies on DT-μCF. *Hyperfine Interact.* **1996**, *101*, 521–538. [CrossRef]
4. TOYO TANSO. Available online: http://www.toyotanso.co.jp/index_en (accessed on 13 June 2017).
5. Makimura, S.; Miyake, Y.; Kawamura, N.; Strasser, P.; Koda, A.; Shimomura, K.; Fujimori, H.; Nishiyama, K.; Kato, M.; Nakahara, K.; et al. Present status of construction for the muon target in J-PARC. *Nucl. Instrum. Meth. A* **2009**, *600*, 146–149. [CrossRef]

6. Makimura, S.; Shimizu, R.; Kawamura, N.; Miyake, Y.; Kobayashi, Y.; Koda, A.; Kato, M.; Fujimori, H.; Shimomura, K.; Strasser, P.; et al. Report for muon production target with proton beam operation in J-PARC. In Proceedings of the 8th Annual Meeting of Particle Accelerator Society of Japan, Tsukuba, Japan, 1–3 August 2011; Particle Accelerator Society of Japan: Tokyo, Japan, 2011; 45008267, pp. 1188–1191.

7. Matsuo, H. Irradiation damage in nuclear graphite and carbon material changes in dimension and physical properties caused by neutron irradiation and heat treatment. *Graphite* **1991**, *1991*, 290–302. [CrossRef]

8. Makimura, S.; Kawamura, N.; Kojima, K.M.; Koda, A.; Kurosawa, N.; Shimizu, R.; Strasser, P.; Nakahara, J.; Miyake, Y. Remote-controlled non-destructive measurement for thermal conductivity of highly radioactive isotropic graphite used as the muon production target at J-PARC/MUSE. *J. Nucl. Mater.* **2014**, *450*, 110–116. [CrossRef]

9. Makimura, S.; Kawamura, N.; Kato, M.; Kobayashi, Y.; Miyake, Y.; Koda, A.; Fujimori, H.; Shimomura, K.; Strasser, P.; Nishiyama, K.; et al. Remotely controlled replacement of highly radioactive components in J-PARC/MUSE. In Proceedings of the 7th Annual Meeting of Particle Accelerator Society of Japan, Himeji, Japan, 4–6 August 2010; Particle Accelerator Society of Japan: Tokyo, Japan, 2010; 44081307, pp. 479–483.

10. Makimura, S.; Kawamura, N.; Onizawa, S.; Matsuzawa, Y.; Tabe, M.; Kobayashi, Y.; Shimizu, R.; Taniguchi, Y.; Fujimori, H.; Ikedo, Y.; et al. Development of muon rotating target at J-PARC/MUSE. *J. Radioanal. Nucl. Chem.* **2015**, *305*, 811–815. [CrossRef]

11. Iwase, H.; Niita, T. Development of general-purpose particle and heavy ion transport Monte Carlo code. *J. Nucl. Sci. Technol.* **2002**, *39*, 1142–1151. [CrossRef]

12. Kawamura, N.; Makimura, M.; Strasser, P.; Koda, A.; Fujimori, H.; Nishiyama, K.; Miyake, Y. Design strategy for devices under high radiation field in J-PARC muon facility. *Nucl. Instrum. Meth. A* **2009**, *600*, 114–116. [CrossRef]

13. Makimura, S.; Ozaki, H.; Okamura, H.; Futakawa, M.; Naoe, T.; Miyake, Y.; Kawamura, N.; Nishiyama, K.; Kawai, M. The present status of R&D for the muon target at J-PARC: The development of silver-brazing method for graphite. *J. Nucl. Mater.* **2008**, *377*, 28–33. [CrossRef]

14. Makimura, S.; Kawamura, N.; Onizawa, S.; Matsuzawa, Y.; Tabe, M.; Kobayashi, Y.; Shimizu, R.; Fujimori, H.; Ikedo, Y.; Kadono, R.; et al. Present status of muon production target at J-PARC/MUSE. *JPS Conf. Proc.* **2015**, *8*, 051002. [CrossRef]

15. Makimura, S.; Matoba, S.; Kawamura, N.; Onizawa, S.; Matsuzawa, Y.; Tabe, M.; Kobayashi, Y.; Fujimori, H.; Ikedo, Y.; Koda, A.; et al. Present status of muon rotating target at J-PARC/MUSE. In Proceedings of the 12th Annual Meeting of Particle Accelerator Society of Japan, Tsuruga, Japan, 5–7 August 2015; Particle Accelerator Society of Japan: Tokyo, Japan, 2015; Volume FROM13, pp. 261–264.

16. JTEKT. Available online: http://www.jtekt.co.jp/e/index.html (accessed on 13 June 2017).

17. Bakule, P.; Matsuda, Y.; Miyake, Y.; Strasser, P.; Shimomura, K.; Makimura, S.; Nagamine, K. Slow muon experiment by laser resonant ionization method at RIKEN-RAL muon facility. *Spectrochim. Acta B* **2003**, *58*, 1019–1030. [CrossRef]

18. Strasser, P.; Shimomura, K.; Koda, A.; Kawamura, N.; Fujimori, H.; Makimura, S.; Kobayashi, Y.; Nakahara, K.; Kato, M.; Takeshita, S.; et al. J-PARC decay muon channel construction status. *J. Phys. Conf. Ser.* **2010**, *225*, 012050. [CrossRef]

19. Strasser, P.; Fujimori, H.; Koseki, K.; Hori, K.; Matsumoto, H.; Shimomura, K.; Koda, A.; Kawamura, N.; Makimura, S.; Kato, M.; et al. New muon kicker system for the decay muon beamline at J-PARC. *Phys. Procedia* **2012**, *30*, 65–68. [CrossRef]

20. Miyake, Y.; Ikedo, Y.; Shimomura, K.; Strasser, P.; Nagatomo, T.; Nakamura, J.; Makimura, S.; Kawamura, N.; Fujimori, H.; Koda, A.; et al. Ultra slow muon project at J-PARC, MUSE. *JPS Conf. Proc.* **2014**, *2*, 010101. [CrossRef]

21. Nagamine, K.; Miyake, Y.; Shimomura, K.; Birrer, P.; Marangos, J.P.; Iwasaki, M.; Strasser, P.; Kuga, T. Ultraslow positive-muon generation by laser ionization of thermal muonium from hot tungsten at primary proton beam. *Phys. Rev. Lett.* **1995**, *74*, 4811–4814. [CrossRef] [PubMed]

22. Ikedo, Y.; Miyake, Y.; Shimomura, K.; Strasser, P.; Nishiyama, K.; Kawamura, N.; Fujimori, H.; Makimura, S.; Koda, A.; Ogitsu, T.; et al. Status of the Superomega muon beam line at J-PARC. *Phys. Procedia* **2012**, *30*, 34–37. [CrossRef]

23. Ikedo, Y.; Miyake, Y.; Shimomura, K.; Strasser, P.; Kawamura, N.; Nishiyama, K.; Makimura, S.; Fujimori, H.; Koda, A.; Nakamura, J.; et al. Positron separators in Superomega muon beamline at J-PARC. *Nucl. Instrum. Meth. B* **2013**, *317*, 365–368. [CrossRef]

24. Strasser, P.; Ikedo, Y.; Miyake, Y.; Shimomura, K.; Kawamura, N.; Nishiyama, K.; Makimura, S.; Fujimori, H.; Koda, A.; Nakamura, J.; et al. Superconducting curved transport solenoid with dipole coils for charge selection of the muon beam. *Nucl. Instrum. Meth. B* **2013**, *317*, 361–364. [CrossRef]

25. Oishi, Y.; Okamura, K.; Miyazaki, K.; Saito, N.; Iwasaki, M.; Wada, S. All-solid-state laser amplifiers for intense Lyman-α generation. *JPS Conf. Proc.* **2014**, *2*, 010105. [CrossRef]

26. Nakamura, J.; Oishi, Y.; Saito, N.; Miyazaki, K.; Okamura, K.; Higemoto, W.; Ikedo, Y.; Kojima, K.M.; Strasser, P.; Nagatomo, T.; et al. Optimal crossed overlap of coherent vacuum ultraviolet radiation and thermal muonium emission for μSR with the Ultra Slow Muon. *J. Phys. Conf. Ser.* **2015**, *551*, 012066. [CrossRef]

27. Strasser, P.; Ikedo, Y.; Makimura, S.; Nakamura, J.; Nishiyama, K.; Shimomura, K.; Fujimori, H.; Adachi, T.; Koda, A.; Kawamura, N.; et al. Design and construction of the ultra-slow muon beamline at J-PARC/MUSE. *J. Phys. Conf. Ser.* **2015**, *551*, 012065. [CrossRef]

28. Shimomura, K. Possibility of precise measurement of muonium HFS at J-PARC MUSE. *AIP Conf. Proc.* **2011**, *1382*, 245–247. [CrossRef]

29. Aoki, M.; DeeMe Collaboration. An experimental search for muon-electron conversion in nuclear field at sensitivity of 10^{-14} with a pulsed proton beam. *AIP Conf. Proc.* **2012**, *1441*, 599–601. [CrossRef]

30. Saito, N.; J-PARC g−2/EDM Collaboration. A novel precision measurement of muon g−2 and EDM at J-PARC. *AIP Conf. Proc.* **2012**, *1467*, 45–56. [CrossRef]

31. G4beamline. Available online: http://public.muonsinc.com/ (accessed on 13 June 2017).

32. Higemoto, W.; Ito, T.U.; Ninomiya, K.; Heffner, R.H.; Shimomura, K.; Nishiyama, K.; Miyake, Y. Muon beam slicer at J-PARC MUSE. *Phys. Procedia* **2012**, *30*, 30–33. [CrossRef]

33. Hiraishi, M.; Iimura, S.; Kojima, K.M.; Yamaura, J.; Hiraka, H.; Ikeda, K.; Miao, P.; Ishikawa, Y.; Torii, S.; Miyazaki, M.; et al. Bipartite magnetic parent phases in the iron oxypnictide superconductor. *Nat. Phys.* **2014**, *10*, 300–303. [CrossRef]

34. Iimura, S.; Matuishi, S.; Sato, H.; Hanna, T.; Muraba, Y.; Kim, S.W.; Kim, J.E.; Takata, M.; Hosono, H. Two-dome structure in electron-doped iron arsenide superconductors. *Nat. Commun.* **2012**, *3*, 943. [CrossRef] [PubMed]

35. Fujiwara, N.; Tsutsumi, S.; Iimura, S.; Matsuishi, S.; Hosono, H.; Yamakawa, Y.; Kontani, H. Detection of antiferromagnetic ordering in heavily doped LaFeAsO$_{1-x}$H$_x$ pnictide superconductors using nuclear-magnetic-resonance techniques. *Phys. Rev. Lett.* **2013**, *111*, 097002. [CrossRef] [PubMed]

36. Kojima, K.M.; Kadono, R.; Miyazaki, M.; Hiraishi, M.; Yamauchi, I.; Koda, A.; Tsuchiya, Y.; Suzuki, H.S.; Kitazawa, H. Magnetic frustration in iridium spinel compound CuIr$_2$S$_4$. *Phys. Rev. Lett.* **2014**, *112*, 087203. [CrossRef]

37. Ito, T.U.; Higemoto, W.; Matsuda, T.D.; Koda, A.; Shimomura, K. Shallow donor level associated with hydrogen impurities in undoped BaTiO$_3$. *Appl. Phys. Lett.* **2013**, *103*, 042905. [CrossRef]

38. Terada, K.; Ninomiya, N.; Osawa, T.; Tachibana, S.; Miyake, Y.; Kubo, M.K.; Kawamura, N.; Higemoto, W.; Tsuchiyama, A.; Ebihara, M.; et al. A new X-ray fluorescence spectroscopy for extraterrestrial materials using a muon beam. *Sci. Rep.* **2014**, *4*, 5072. [CrossRef] [PubMed]

MDPI

St. Alban-Anlage 66

4052 Basel

Switzerland

Tel. +41 61 683 77 34

Fax +41 61 302 89 18

www.mdpi.com

Quantum Beam Science Editorial Office

E-mail: qubs@mdpi.com

www.mdpi.com/journal/qubs

www.ingramcontent.com/pod-product-compliance
Lightning Source LLC
Chambersburg PA
CBHW051904210326
41597CB00033B/6014